Geomechanics for Energy and a Sustainable Environment

Geomechanics for Energy and a Sustainable Environment

Special Issue Editors

Gye-Chun Cho
Ilhan Chang

MDPI • Basel • Beijing • Wuhan • Barcelona • Belgrade

MDPI

Special Issue Editors

Gye-Chun Cho
Korea Advanced Institute of Science and Technology
Korea

Ilhan Chang
University of New South Wales
Australia

Editorial Office
MDPI
St. Alban-Anlage 66
4052 Basel, Switzerland

This is a reprint of articles from the Special Issue published online in the open access journal *Energies* (ISSN 1996-1073) from 2018 to 2019 (available at: https://www.mdpi.com/journal/energies/special_issues/geomechanics_energy).

For citation purposes, cite each article independently as indicated on the article page online and as indicated below:

LastName, A.A.; LastName, B.B.; LastName, C.C. Article Title. *Journal Name* **Year**, *Article Number*, Page Range.

ISBN 978-3-03928-150-3 (Pbk)
ISBN 978-3-03928-151-0 (PDF)

Contents

About the Special Issue Editors

Gye-Chun Cho, Ph.D., professor in the Department of Civil and Environmental Engineering in the Korea Advanced Institute of Science and Technology (KAIST), received his B.S. and M.S. from Korea University (South Korea) and his Ph.D. from the Georgia Institute of Technology (USA). His main interests are geotechnical engineering, energy geotechnology, tunneling and rock excavation, methane and CO_2 embedded hydrates, biopolymer-based soil treatment, and construction machinery and automation. He has published more than 400 papers in peer-reviewed journals, books, and conference proceedings.

Ilhan Chang Ph.D. is an academic Senior Lecturer at the School and Engineering and Information Technology, University of New South Wales (Australia). He received his B.S., M.S., and Ph.D. degrees in Civil and Geotechnical Engineering from the Korea Advanced Institute of Science and Technology (KAIST). He gained research and academic experience as a Research Assistant Professor (KAIST; 2011), a Senior Researcher (Korea Institute of Civil Engineering and Building Technology; 2012–2016), and a Senior Lecturer (UNSW; 2017–present). His research activities include soil characterization, micro geomechanics, ground improvement, soil erosion and preservation, and multi-disciplinary convergences such as bio-soils. As a result of his research, he has published more than 140 papers and patents, including 35 peer-reviewed SCI(E) journal articles.

Preface to "Geomechanics for Energy and a Sustainable Environment"

Energy consumption is strongly linked to the development and quality of human civilization. On average, 85% of our primary energy comes from fossil fuels; as a result, carbon-based resources face depletion, and there are concerns about massive greenhouse gas emission and corresponding climate change. Energy geotechnology must play a key role in the development of a sustainable energy scheme, involving energy resources such as natural carbohydrates, nuclear energy, and renewable sources (wind, solar, geothermal, hydropower, biofuels, and tidal and wave energy), but it must not be restricted. Moreover, geotechnical engineering is required to provide new techniques to preserve the environment from a sustainability perspective, including CO_2 emission and energy-related waste (e.g., bottom and fly ashes, and nuclear waste) reduction, as well as the introduction of new, environmentally-friendly, and low-carbon-emitting materials for sustainable development.

This book includes recent advances in *Geomechanics for Energy and the Sustainable Environment*, including four research articles and one state-of-the-art review. We hope this book can provide guidance on how to deal with conventional and renewable energy sources and environment-related geotechnical engineering issues, from fundamental research to practical implementation. In addition, recent attempts in CO_2 and industrial waste deduction, and the development of new, bio-inspired materials/methods for sustainable development are introduced in this book.

.

Gye-Chun Cho, Ilhan Chang
Special Issue Editors

energies

MDPI

Article

Lattice Boltzmann Simulation for the Forming Process of Artificial Frozen Soil Wall

Linfang Shen, Zhiliang Wang *, Pengyu Wang and Libin Xin

Faculty of Civil Engineering and Mechanics, Kunming University of Science and Technology, Kunming 650500, China; shenlinfang@kmust.edu.cn (L.S.); wangpengyu18@126.com (P.W.); 18904722465@163.com (L.X.)
* Correspondence: wangzhiliang@kmust.edu.cn

Received: 25 October 2018; Accepted: 19 December 2018; Published: 24 December 2018

Abstract: A lattice Boltzmann model is proposed to simulate the forming process of artificial frozen soil wall. The enthalpy method is applied to deal with the latent-heat source term, and the adjustable thermal diffusivity is utilized to handle the change of thermophysical parameters. The model is tested by the heat conduction with solid–liquid phase change in semi-infinite space, which shows a good consistence between the numerical and analytical solutions, and the mesh resolution has little effect on the numerical results. Lastly, the development of frozen soil wall is discussed when the freezing pipes are arranged in a square. The results show that the evolution of temperature field with time is closely related to the distance from the freezing pipe. For the soil near freezing pipe, the temperature gradient is larger, the soil temperature drops rapidly and freezes in a short time. The time history curve of temperature is relatively smooth. For the soil far away from freezing pipe, the temperature evolution curve has obvious multistage, which can be divided into four stages: cooling, phase change, partly frozen and completely frozen. The spacing of freezing pipes has a significant influence on the overlapping time of artificial frozen soil wall, and there is a power function relationship between them.

Keywords: lattice Boltzmann method; artificial frozen soil wall; temperature field; phase change; numerical simulation

---

1. Introduction

As an effective temporary ground improvement technique, artificial ground freezing (AGF) has been widely adopted in geotechnical engineering [1,2], including departure and reception of shield tunnel, tunnels connecting passage in Metro, mine shaft sinking and municipal engineering, etc. It has advantages of strong stratum adaptability, good sealing performance, high strength and little influence on the surrounding environment. In ground improvement engineering, the development of the frozen soil wall determines its average temperature, thickness, physical and mechanical properties. These indexes reflect the strength and stability of the frozen soil wall, which is directly related to the scheduling management of projects. In the process of artificial freezing, the temperature evolution of soil is a transient heat conduction problem including ice–water phase change, latent heat release, internal heat source, moving boundary and irregular geometric boundary. The soil temperature distribution is also affected by the interaction of freezing pipes. Therefore, the forming process of frozen soil wall is very complicated, and a better understanding of heat transfer mechanisms is essentially important, which can provide necessary technical guarantees for the implementation of artificial freezing projects.

Based on the steady heat conduction theory, a large number of scholars have proposed the analytical methods of temperature evolution with a single pipe, single or double row pipes [3–5]. However, these methods are limited to simplified boundary conditions and idealized initial conditions.

Besides, the theoretical formula is too complicated for engineering application. In recent years, numerical methods have been widely applied in heat conduction problems with phase change. Singh [6] studied the flow and heat transfer characteristics of a phase transition, melting problem. Santos [7] applied the finite element method to predict freezing times of mushrooms. Farrokhpanah [8] introduced a new smoothed particle hydrodynamics (SPH) method to model the heat transfer with phase change considering the latent heat released (absorbed) during solidification (melting). Furenes [9] used the event location algorithm in the finite difference method for phase-change problems.

When compared with the traditional numerical methods, the lattice Boltzmann method (LBM) enjoys advantages of both macroscopic and microscopic approaches [10,11]. It has clear physical conception, easy programming, high computational efficiency and is easy to apply for complex domains [12–14]. So, LBM has been explored to deal with heat conduction problems with phase change. Miller [15] proposed a simple model for the liquid-solid phase change based on the lattice Boltzmann method with enhanced collisions. Jiaung [16] firstly developed an enthalpy-based lattice Boltzmann model for simulating solid/liquid phase change problem governed by the heat conduction equation. Huber et al. [17–19] improved this model, and used it to couple thermal convection and phase change of single-component systems. Eshraghi [20] developed a new variation to solve the heat conduction with phase change by treating implicitly the latent heat source term. Huang [21] proposed a new lattice Boltzmann model to treat the latent heat source term by modifying the equilibrium distribution function. Sadeghi [22,23] proposed a three-dimensional Boltzmann model to study the film-boiling phenomenon. Chatterjee [24] extended the lattice Boltzmann formulation to simulate three-dimensional heat diffusion coupled with solid–liquid phase change. Li [25] presented a three-dimensional multiple-relaxation-time lattice Boltzmann model for the solid–liquid phase change based on the enthalpy conservation equation. The above studies mainly neglected the change of thermal diffusivity for simplifying the calculation. However, the thermal diffusivity of liquid water is only 1/9 of ice, and the water content of artificial frozen soil is generally high, so it is necessary to consider the change of thermal diffusivity during the forming process of artificial frozen soil wall.

In this paper, the enthalpy approach is applied to treat the latent heat source term in the energy equation, the adjustable thermal diffusivity is utilized to simulate the change of thermophysical parameters, and a thermal lattice Boltzmann model is proposed to simulate the forming process of artificial frozen soil wall. The model is subsequently tested with the solid–liquid phase change of pure substance in semi-infinite space. Finally, the forming process of an artificial frozen wall is simulated when the four freezing pipes are arranged in a square, the evolution of the freezing front and the temperature distribution are analyzed during the artificial freezing process, which provides the theoretical basis for the design and construction of practical engineering.

2. Heat Conduction Model of Soil Freezing

2.1. Assumptions

To develop the heat conduction model with phase change for soil, the following assumptions are made:

(1) Ignoring the influence of pore structure, soil is regard as the continuous, homogeneous and isotropic medium.

(2) According to the physical state of water, the freezing soil is divided into two parts: solid and liquid zones, and the thermophysical parameters are constant in each zone.

(3) The freezing temperature of soil is constant, and the liquid phase fraction is applied to trace the solid–liquid interface.

(4) The coupling of temperature, stress and moisture is neglected during the artificial freezing process.

2.2. Mathematical Model

Compared with the three-dimensional model, the two-dimensional one could reflect the general evolution laws and save a lot of computing resources, so it is selected in the present paper. According to energy conservation, the mathematical model of soil freezing can be expressed as [26]:

Solid phase

$$\frac{\partial T_s}{\partial t} = \alpha_s \left[\frac{\partial^2 T_s}{\partial x^2} + \frac{\partial^2 T_s}{\partial y^2} \right],$$ (1)

Liquid phase

$$\frac{\partial T_l}{\partial t} = \alpha_l \left[\frac{\partial^2 T_l}{\partial x^2} + \frac{\partial^2 T_l}{\partial y^2} \right],$$ (2)

The solid–liquid interface

$$\begin{cases} T_s = T_l \\ k_s \frac{\partial T_s}{\partial r} - k_l \frac{\partial T_l}{\partial r} = \rho_s L_a \frac{\partial S}{\partial t} \end{cases},$$ (3)

where T is the temperature, t is the time. ρ, k and α are the density, thermal conductivity and thermal diffusivity, respectively, and all these parameters can be determined by the volume fraction of each component in soil [27,28]. L_a is the latent heat, S is the position of solid–liquid interface. r is the normal direction of frozen front. The subscript l represents that the water in the soil is the liquid phase, and s the solid phase.

2.3. Phase Change Treatment

The difficulties of temperature prediction lie in the treatment of latent heat and the movement of solid–liquid interface during the soil freezing process. In this paper, the enthalpy model proposed by Shamsundar [29] is adopt to develop the unified energy equation in the whole region (including liquid zone, solid zone and solid–liquid interface), and the solid–liquid interface is determined by solving the enthalpy parameter. This model does not need to separate the solid and liquid phases, and trace the solid–liquid interface. Mathematically, it has been proved to be equivalent to the heat conduction equation with phase change [30]. So, the Equations (1)–(3) can be unified as

$$\rho \frac{\partial H}{\partial t} = k \left(\frac{\partial^2 T}{\partial x^2} + \frac{\partial^2 T}{\partial y^2} \right),$$ (4)

where H is the total enthalpy, which can be divided into sensible and latent enthalpy components as:

$$H = C_p T + \varphi L_a,$$ (5)

where C_p is the specific heat capacity, φ is the liquid phase fraction, which is 0 for solid zone, 1 for liquid zone.

When the freezing temperature T_f is constant, the relationship between the liquid phase fraction φ and the total enthalpy H can be expressed as the follows:

$$\varphi = \begin{cases} 0 & H < C_p T_f \\ \frac{H - C_p T_f}{L_a} & C_p T_f \le H \le C_p T_f + L_a \\ 1 & H > C_p T_f + L_a \end{cases},$$ (6)

Substituting Equation (5) into Equation (4) yields:

$$\frac{\partial T}{\partial t} = \alpha \left(\frac{\partial^2 T}{\partial x^2} + \frac{\partial^2 T}{\partial y^2} \right) - \frac{L_a}{C_p} \frac{\partial \varphi}{\partial t},$$ (7)

Equation (7) is applied to describe the evolution of temperature field during the soil freezing process. In terms of the liquid phase fraction φ, the thermal diffusivity α can be expressed as:

$$\alpha = \varphi\alpha_l + (1 - \varphi)\alpha_s, \tag{8}$$

2.4. Dimensionless Treatment

To facilitate transformation between physical and lattice units, the following dimensionless parameters are introduced

$$X = \frac{x}{L}, \theta = \frac{T - T_0}{T_i - T_0}, F_0 = \frac{\alpha_s t}{L^2}, Ste = \frac{C_p\left(T_f - T_0\right)}{L_a}, \tag{9}$$

where X, θ are the dimensionless coordinate and temperature. Ste, F_0 are the Stefan number and Fourier number, respectively. L is the reference length, T_0 is the temperature of cold source, T_i is the initial temperature of soil.

3. Lattice Boltzmann Model

3.1. Lattice Boltzmann Equation

The d-dimensional m-speed (DdQm) model proposed by Qian et al. [31], is the basic model of LBM. Compared with D2Q9 model, the D2Q4 model (Figure 1) has not only comparable accuracy but also better computational efficiency for temperature evolution, so it is employed to solve the heat conduction equation, which is governed by Equation (7). And the enthalpy model is adopted to treat the latent heat source term. The discrete form of the lattice Boltzmann equation can be written as:

$$g_i(\mathbf{r} + \mathbf{e_i}\delta_t, t + \delta_t) - g_i(\mathbf{r}, t) = -\frac{g_i(\mathbf{r}, t) - g_i^{eq}(\mathbf{r}, t)}{\tau} + \omega_i Sr\delta_t \qquad (i = 0, 1, 2, 3), \tag{10}$$

where $g_i(\mathbf{r}, t)$ is the temperature distribution function of the ith direction at the lattice site $\mathbf{r}$ and time t, $g_i^{eq}(\mathbf{r}, t)$ represents the equilibrium distribution function, τ is the dimensionless relaxation time, whose value should insure to be within (0.5,2) [32], $\mathbf{e_i}$ is the discrete velocity in the lattice, which is composed of the velocity vectors:

$$\mathbf{e} = c\begin{bmatrix} 1 & 0 & -1 & 0 \\ 0 & 1 & 0 & -1 \end{bmatrix}, \tag{11}$$

where c is the lattice speed, and $c = \delta_x/\delta_t$, δ_x, δ_t are the lattice space and time step, respectively, and they are usually taken as $\delta_x = \delta_t = 1$ for simplifying the calculation.

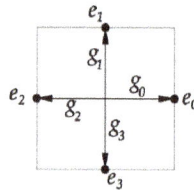

Figure 1. D2Q4 model.

$g_i^{eq}(\mathbf{r}, t)$ can be described as

$$g_i^{eq}(\mathbf{r}, t) = \omega_i T(\mathbf{r}, t), \tag{12}$$

where ω_i is the weight factor in the ith direction, and $\omega_0 = \omega_1 = \omega_2 = \omega_3 = 1/4$ for D2Q4 model.

Sr is the heat source term, according to Equation (7), it can be expressed as:

$$Sr = -\frac{L_a}{C_p}\frac{\partial \varphi}{\partial t},$$ (13)

Based on the first-order expansion, the heat source term Sr is approximately discretized into:

$$Sr = -\frac{L_a}{C_p}\frac{[\varphi(t+\delta_t) - \varphi(t)]}{\delta_t},$$ (14)

By Chapman–Enskog expansion, Equation (10) can be recovered to the macroscopic heat conduction Equation (7), the thermal diffusivity α is given by

$$\alpha = c_s^2\left(\tau - \frac{1}{2}\right)\delta_t,$$ (15)

where c_s is lattice sound speed, for the D2Q4 model, $c_s^2 = c^2/2$.

The macroscopic temperature can be calculated as:

$$T(\mathbf{r},t) = \sum_{i=0}^{3} g_i(\mathbf{r},t),$$ (16)

3.2. Boundary Conditions

The adiabatic boundary involved in the present study, is handled by the non-equilibrium extrapolation scheme, proposed by Guo et al. [33] in 2002, which has second order accuracy. The main idea of this approach is to divide the temperature distribution functions at the boundary node N_B into its equilibrium and non-equilibrium parts,

$$g_i(N_B,t) = g_i^{eq}(N_B,t) + g_i^{neq}(N_B,t),$$ (17)

The equilibrium part $g_i^{eq}(N_B,t)$ can be got by Equation (12), and the non-equilibrium part $g_i^{neq}(N_B,t)$ is approximated by extrapolating from the neighboring node N_O,

$$g_i^{neq}(N_B,t) = g_i(N_O,t) - g_i^{eq}(N_O,t),$$ (18)

3.3. Unit Conversion

For numerical methods, it is necessary to achieve the unit conversion between physical and lattice units. In the present study, the dimensionless treatment is carried out for all the parameters, which ensures the consistency of heat transfer criterion. The non-dimensional numbers such as Stefan number and Fourier number, are used as a bridge to realize the conversion between two unit systems.

For the thermophysical parameters (latent heat L_a, specific heat C_p), the unit conversion can be achieved based on the Stefan number Ste.

$$Ste = \frac{C_{pp}\left(T_{fp} - T_{0p}\right)}{L_{ap}} = \frac{C_{pL}\left(T_{fL} - T_{0L}\right)}{L_{aL}},$$ (19)

where subscript p represents the physical unit, and L the lattice unit.

According to an artificial freezing project, the sandy silt [34] is taken as the research object. The unit conversion of thermophysical parameters is handled, and the comparison between physical and lattice units is shown in Table 1.

Table 1. Comparisons of thermophysical parameters between physical and lattice units.

| Unit | Latent Heat L_a | Thermal Diffusivity α | | Heat Capacity C_p | Initial Temperature T_i | Freezing Temperature T_f | Temperature of Cold Source T_0 |
		Solid Phase α_s	Liquid Phase α_l				
Physical unit	121.09 kJ/kg	5.97×10^{-7} m^2/s	4.86×10^{-7} m^2/s	1.449 kJ/(kg·°C)	10 °C	0 °C	−30 °C
Lattice unit	1.0	0.125	0.10176	0.47865	1.0	0.75	0.0

For the physical time t_p and lattice time steps N, the relationship between them can be established according to the Fourier number F_o.

$$F_o = \frac{\alpha_p t_p}{L_p^2} = \frac{\alpha_L t_L}{L_L^2} \tag{20}$$

where t_p and t_L are the time in physical and lattice unit, $t_L = N\delta_t$. L_p and L_L are the reference length in physical and lattice units, if L_p is the length of calculation domain, then $L_L = n\delta_x$, n is the corresponding number of lattices.

If the physical model of 4 m × 4 m is divided into a lattice of 1000 × 1000 grid cells, and the values of α_p and α_L are assigned according to Table 1. The relationship between t_p and t_L can be deduced as

$$t_p = \frac{\alpha_L L_p^2}{\alpha_p L_L^2} t_L = \frac{0.125 \times 4^2}{5.97 \times 10^{-7} \times 1000^2} t_L = 3.35 t_L, \tag{21}$$

It can be known that if δ_t is selected as 1, the physical time 1 s is corresponding to 3.35-time steps and 1 day to 25,790-time steps.

3.4. Flowchart of Program Realization

Considering the effects of heat transfer, latent heat and movement of phase-change interface during the soil freezing process, the lattice Boltzmann model is proposed based on the enthalpy method, the corresponding flowchart is shown in Figure 2.

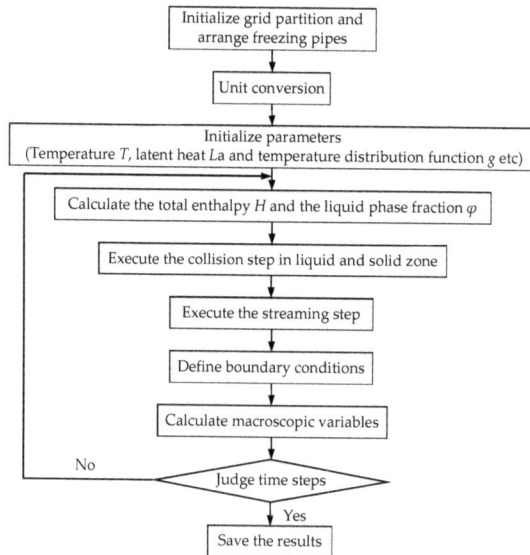

Figure 2. Flowchart of program realization for soil freezing.

3.5. Verification

To verify accuracy of proposed model, LBM is applied to simulate the heat conduction problem with solid–liquid phase change in semi-infinite space, as shown in Figure 3. At the time $t = 0$, the uniform initial temperature is T_i, and the substance is in the liquid state. The cold source temperature T_0 is set at the position of $x = 0$, and keeps constant at $t > 0$. The analytical solution of the solid–liquid interface $S(t)$ and temperature $T(x, t)$ are as follows [26]

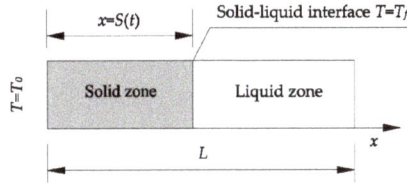

Figure 3. Diagram of heat conduction problem with phase change in semi-infinite space.

Solid–liquid interface

$$S(t) = 2\lambda\sqrt{\alpha_s t}, \tag{22}$$

Solid phase

$$T_s(x, t) = T_0 + \frac{\left(T_f - T_0\right)}{erf(\lambda)} erf\left(\frac{x}{2\sqrt{\alpha_s t}}\right), \tag{23}$$

Liquid phase

$$T_l(x, t) = T_i + \frac{\left(T_f - T_i\right)}{erfc\left(\lambda\sqrt{\frac{\alpha_s}{\alpha_l}}\right)} erfc\left(\frac{x}{2\sqrt{\alpha_l t}}\right), \tag{24}$$

where λ is the unknown parameter, which can be obtained by the following transcendental equation.

$$\frac{e^{-\lambda^2}}{erf(\lambda)} + \frac{k_l}{k_s}\left(\frac{\alpha_s}{\alpha_l}\right)^{\frac{1}{2}} \frac{\left(T_f - T_i\right)}{\left(T_f - T_0\right)} \frac{e^{-\lambda^2\left(\frac{\alpha_s}{\alpha_l}\right)}}{erfc\left(\lambda\sqrt{\frac{\alpha_s}{\alpha_l}}\right)} = \frac{\lambda L_a\sqrt{\pi}}{C_p\left(T_f - T_0\right)}, \tag{25}$$

In this case, the entire domain has a size of $L \times H = 2\ \text{m} \times 0.16\ \text{m}$, and it is discretized using 125×10, 250×20 and 500×40 grid cells with lattice resolution of 16 mm, 8 mm and 4 mm, respectively. The thermophysical parameters are set according to Table 1. The temperature is fixed at T_0 on the left side, and the right boundary is adiabatic. The lattice spaces are equal in horizontal and vertical direction, $\delta_x = \delta_y = 1.0$ and the time step is $\delta_t = 1.0$.

Figure 4 shows the evolution of solid–liquid interface when $L \times H = 125 \times 10$, 250×20 and 500×40, respectively. It can be seen that the results in the present study are in good agreement with the analytical ones, and the mesh resolution has little effect on the numerical results. So the proposed model can accurately simulate the movement of solid–liquid interface during the freezing process. The temperature and error distribution are shown in Figure 5 after freezing 10 days. The slope change of temperature distribution is observed at the solid–liquid interface, which is induced by the variety of thermal diffusivity. A good consistence can also be seen between the numerical results and analytical ones, which indicates the validity of the proposed model in handling the heat conduction problem with phase change. The errors are defined as the numerical solutions of temperature minus the analytical ones. As shown in Figure 5b, the errors fluctuate near the solid–liquid interface, the finer grid resolution has the smaller error, and the farther the distance from the interface, the smaller the error. The maximum errors at the interface are only $-0.2\ °\text{C}$, $0.05\ °\text{C}$ and $0.025\ °\text{C}$ for $L \times H = 125 \times 10$, 250×20 and 500×40, respectively, which are acceptable for engineering application. And the lattice resolution of 4 mm is selected in the following sections.

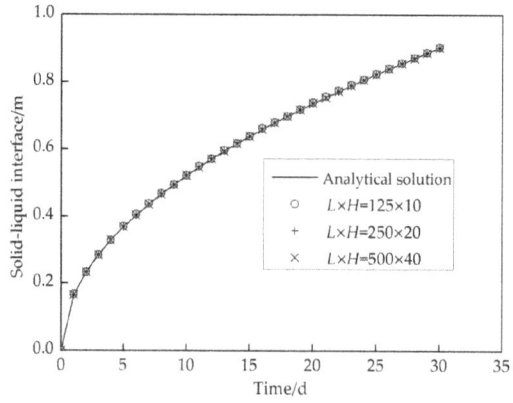

Figure 4. Comparisons between lattice Boltzmann method (LBM) and analytical solutions for the solid–liquid interface.

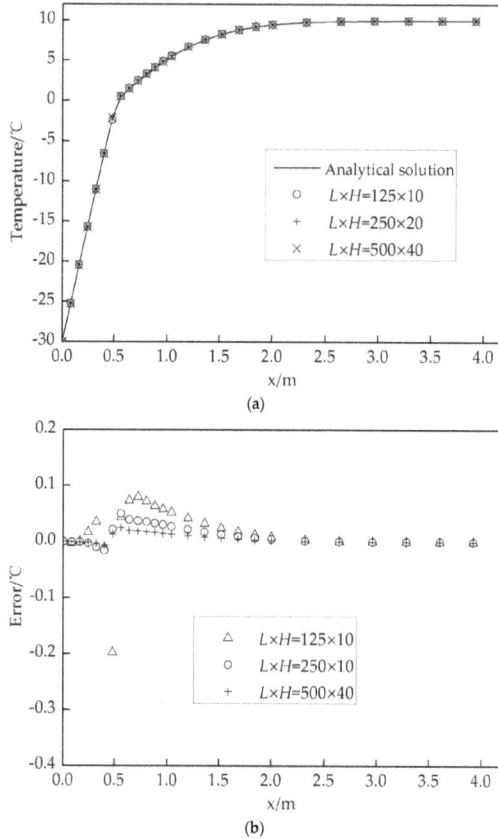

Figure 5. Comparisons between LBM and analytical solutions. (**a**) Temperature distribution; (**b**) Error distribution.

4. Results and Discussion

In practical engineering, the freezing pipes are usually arranged in a rectangular (or diamond) shape. In this paper, the four freezing pipes arranged in a square are selected as an example, which is shown in Figure 6. The development of frozen soil wall and temperature distribution are studied during the freezing process. The dimension of physical model is 4.0 m × 4.0 m. The spacing of freezing pipes is 1.2 m, and the outer diameter is 0.12 m. To ensure the mesh accuracy of the freezing pipes, the entire domain is divided into a lattice of 1000 × 1000 grid cells. The temperature of freezing pipes T_0 is kept at $-30\,^{\circ}$C, the freezing temperature T_f is $0\,^{\circ}$C, and the initial temperature of soil T_i is $10\,^{\circ}$C. The thermophysical parameters of soil are shown in Table 1. The freezing pipes are set as the constant temperature and the four side boundaries of the model are thermally insulated.

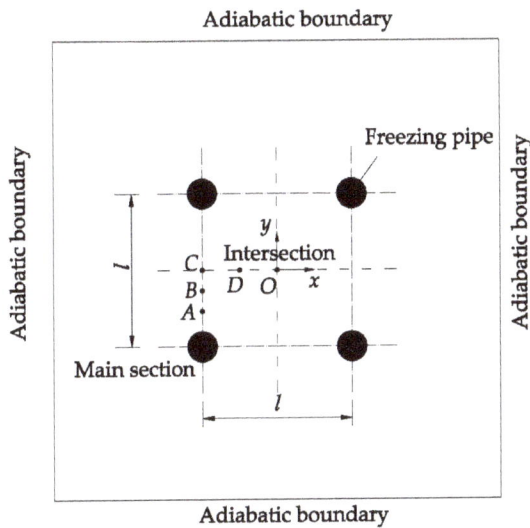

Figure 6. Schematic diagram of the calculation model.

The temporal evolutions of frozen zone are presented in Figure 7, and the time–history curves of temperature at the points *A*, *B*, *C*, *D* and *O* are shown in Figure 8. It can be seen that the temperature at the point *A*, which is closer to the freezing pipe, has larger thermal gradient, the soil freezes quickly, the latent heat has little influence on it, and the time–history curve is smooth. The temperature at the points *C*, *D* and *O*, which is farther from the freezing pipe, has a similar temperature evolution trend. The time–history curves show strong multistage, and for point *C* it can be divided into four stages as shown in Figure 8: (1) Cooling: the temperature drops rapidly at the early stage of artificial freezing, and reaches $0\,^{\circ}$C in about 10 days. (2) Phase change: when the temperature drops to $0\,^{\circ}$C, soil begins to freeze and releases the latent heat. The further away from the freezing pipe, the slower the energy transfers, and the longer the persistent time of phase change stage. (3) Partly frozen: the temperature descends faster in this stage, because there is larger temperature gradient and the thermal diffusivity of frozen soil is greater than that of unfrozen soil. (4) Completely frozen: the temperature evolution is mainly affected by the thermal diffusivity of frozen soil in this stage, the overall trend is relatively stable. For the point *B*, the distance from freezing pipe is moderate, the temperature is somewhere in between, and shows insignificant multistage.

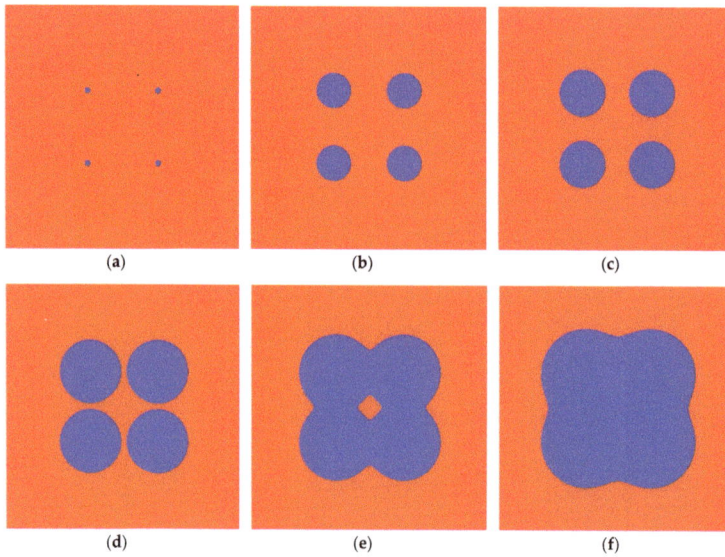

Figure 7. The temporal evolutions of frozen soil wall. (**a**) 0 day; (**b**) 5 day; (**c**) 10 day; (**d**) 20 day; (**e**) 30 day; (**f**) 40 day.

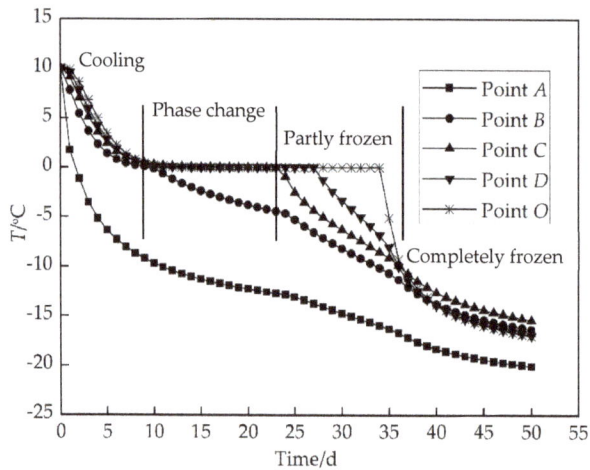

Figure 8. The time–history curves of temperature at the points A, B, C, D and O.

Figure 9 shows the temperature distribution in the main section. Under the action of freezing pipes, the soil temperature decreases rapidly in 10 days, and there is funnel-shaped distribution around the freezing pipes. In about 20 days, the frozen soil wall overlaps in the main section, after that the temperature drops rapidly until soil completely frozen between double rows of freezing pipes in about 40 days, and then the temperature decrease rate slows down gradually.

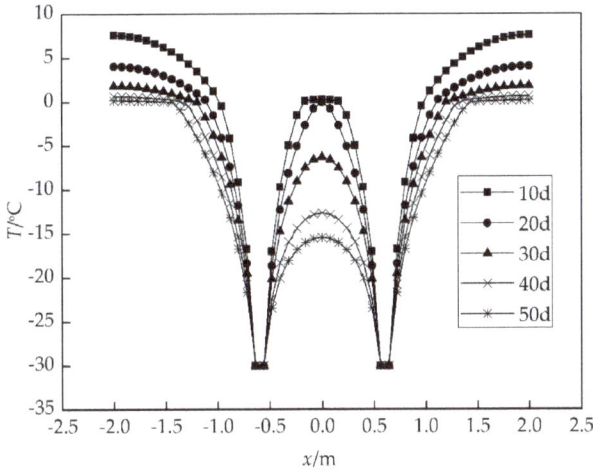

Figure 9. Temperature distribution in the main section.

Figure 10 shows the temperature distribution in the intersection. The distance from the freezing pipes is relatively farther, so the temperature development in the intersection is obviously slower than that in the main section. In about 30 days, most of soil has been frozen in the intersection, then the temperature decreases rapidly, and the stable frozen soil wall forms in about 40 days.

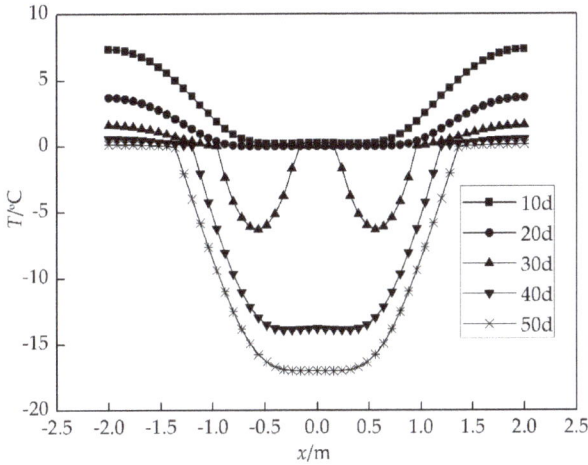

Figure 10. Temperature distribution in the intersection.

The time–history curves of freezing front between two freezing pipes are present in Figure 11, which show that the closer the spacing of freezing pipe is, the faster the freezing front develops, but the whole difference is not significant. Before the frozen soil wall overlapped, the spacing of freezing pipes has little effect on the evolution of freezing front.

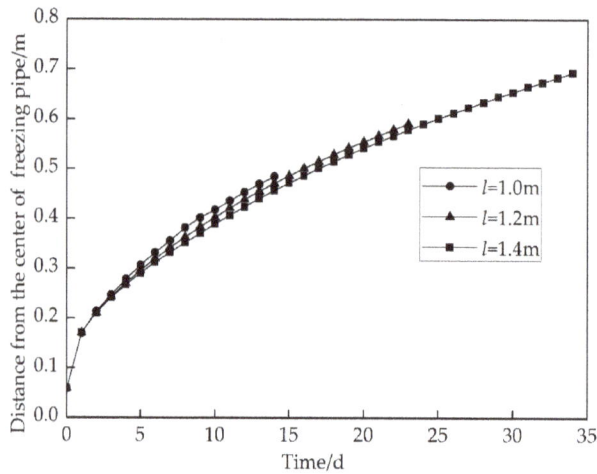

Figure 11. The time–history curve of freezing front.

Figure 12 shows the thickness evolution of frozen soil wall at the point *C* along the *x* direction. The thickness develops faster at the early stage of frozen soil wall overlapped, and the developing speed gradually slows down as time goes on. In general, the thickness evolution has a similar tendency, but for the different freezing pipe spacing, there is a significant difference in overlapping time at the point *O*. When the spacing is 1.0 m, 1.2 m, 1.4 m, respectively, the required time for soil freezing is 8 day, 11 day and 16 day from point *C* to point *O*.

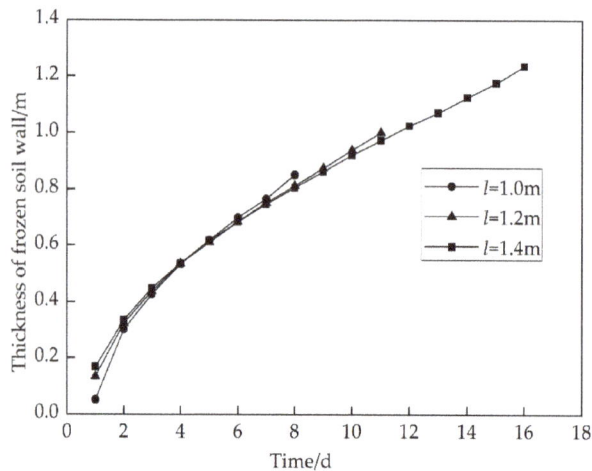

Figure 12. The thickness evolution of frozen soil wall.

Figure 13 shows the relationship between the overlapping time and the spacing of freezing pipes. The overlapping time of frozen soil wall is greatly influenced by the spacing. The freezing times at the point *C* and point *O* increase with the increasing of the spacing, and there is a power function relationship between them. Therefore, at the design stage of artificial freezing engineering, the spacing of freezing pipes should be decided according to the overlapping time of artificial frozen soil wall.

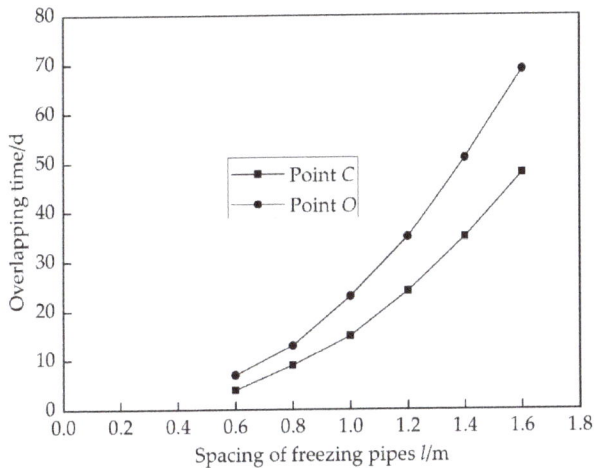

Figure 13. Relationship between the overlapping time and the spacing of freezing pipes.

5. Conclusion

(1) Based on the enthalpy method, a lattice Boltzmann model is proposed to simulate the heat conduction problem with phase change. The model is applied to test the solid–liquid phase change of pure substance, and the results show that the evolution of both temperature distribution and solid–liquid interface are in good agreement with the analytical solutions, and the mesh resolution has little effect on the numerical results.

(2) The temperature evolution of soil is associated with the distance from freezing pipe. When it is closer to the freezing pipe, the time–history curve of temperature is smoother, which is less affected by the latent heat. While it is farther, the time–history curve shows strong multistage, which can be divided into four stages: cooling, phase change, partly frozen, and completely frozen.

(3) Due to the effect of freezing pipes, the soil temperature in the main section decreases rapidly, there is funnel-shaped distribution around the freezing pipes, and the frozen soil wall is overlapped in about 20 days. The temperature development in the intersection is obviously slower compared with that in the main section. In about 30 days, most of soil has been frozen in the intersection, then the temperature decreases rapidly, and the stable frozen soil wall forms in about 40 days.

(4) The spacing of the freezing pipes has a significant influence on the overlapping time of artificial frozen soil wall, and there is a power function relationship between them. But it has little effect on the evolution of freezing front and the thickness of frozen soil wall.

Author Contributions: Each author has made contributions to the present paper. Z.W. proposed this topic and designed the theoretical framework; L.S. wrote the manuscript; P.W. conducted the simulations and analyzed the data; L.X. provided simulation support. All authors have read and approved the final manuscript.

Funding: This work is supported by the National Natural Science Foundation of China (51508253, 51668028) and the Applied Basic Research Project of Yunnan (2016FB077).

Conflicts of Interest: The authors declare no conflict of interest.

References

1. Marwan, A.; Zhou, M.M.; Abdelrehim, M.Z.; Meschke, G. Optimization of artificial ground freezing in tunneling in the presence of seepage flow. *Comput. Geotech.* **2016**, *75*, 112–125. [CrossRef]
2. Pimentel, E.; Sres, A.; Anagnostou, G. Large-scale laboratory tests on artificial ground freezing under seepage-flow conditions. *Géotechnique* **2012**, *62*, 227–241. [CrossRef]

3. Hu, X.D.; Zhang, L.Y. Analytical solution to steady-state temperature field of two freezing pipes with different temperatures. *J. Shanghai Jiaotong Univ. Sci.* **2013**, *18*, 706–711. [CrossRef]

4. Li, F.; Xia, M. Study on analytical solution of temperature field of artificial frozen soil by exponent-integral function. *J. Southeast Univ.* **2004**, *34*, 469–473. (In Chinese)

5. Zhou, Y.; Zhou, G.Q. Analytical solution for temperature field around a single freezing pipe considering unfrozen water. *J. China Coal Soc.* **2012**, *37*, 1649–1653. (In Chinese)

6. Singh, S.; Bhargava, R. Numerical simulation of a phase transition problem with natural convection using hybrid FEM/EFGM technique. *Int. J. Numer. Methods Heat Fluid Flow* **2015**, *25*, 570–592. [CrossRef]

7. Santos, M.V.; Lespinard, A.R. Numerical simulation of mushrooms during freezing using the FEM and an enthalpy: Kirchhoff formulation. *Heat Mass Transf.* **2011**, *47*, 1671–1683. [CrossRef]

8. Farrokhpanah, A.; Bussmann, M.; Mostaghimi, J. New smoothed particle hydrodynamics (SPH) formulation for modeling heat conduction with solidification and melting. *Numer. Heat Transf. Part B Fundam.* **2016**, *71*, 299–312. [CrossRef]

9. Furenes, B.; Lie, B. Using event location in finite-difference methods for phase-change problems. *Numer. Heat Transf. Part B Fundam.* **2006**, *50*, 143–155. [CrossRef]

10. Sukop, M.C., Jr.; Thorne, D.T. *Lattice Boltzmann Modeling: An Introduction for Geoscientists and Engineers*; Springer: Berlin, Germany, 2010; ISBN 9783540279815.

11. Guo, Z.; Shu, C. Lattice Boltzmann method and its applications in engineering. *World Sci.* **2013**. [CrossRef]

12. Mohamad, A.A. *Lattice Boltzmann Method: Fundamentals and Engineering (Applications with Computer Codes)*; Springer: London, UK, 2011; ISBN 978-0-85729-454-8.

13. Chen, L. Numerical Investigation of Multiscale Multiple Physicochemical Coupled Reactive Transport Processes in Energy and Environmental Discipline. Ph.D. Thesis, Xi'an Jiaotong University, Xi'an, China, September 2013.

14. Krüger, T.; Kusumaatmaja, H.; Kuzmin, A.; Shardt, O.; Silva, G.; Viggen, E.M. *The lattice Boltzmann Method: Principles and Practice*; Springer: London, UK, 2017; ISBN 978-3-319-44649-3.

15. Miller, W.; Succi, S.; Mansutti, D. Lattice Boltzmann model for anisotropic liquid-solid phase transition. *Phys. Rev. Lett.* **2001**, *86*, 3578. [CrossRef] [PubMed]

16. Jiaung, W.S.; Ho, J.R.; Kuo, C.P. Lattice Boltzmann Method for the heat conduction problem with phase change. *Numer. Heat Transf. Part B Fundam.* **2001**, *39*, 167–187. [CrossRef]

17. Huber, C.; Parmigiani, A.; Chopard, B.; Manga, M.; Bachmann, O. Lattice Boltzmann model for melting with natural convection. *Int. J. Heat Fluid Flow* **2008**, *29*, 1469–1480. [CrossRef]

18. Chatterjee, D.; Chakraborty, S. An enthalpy-source based lattice Boltzmann model for conduction dominated phase change of pure substances. *Int. J. Therm. Sci.* **2008**, *47*, 552–559. [CrossRef]

19. Huo, Y.; Rao, Z. Lattice Boltzmann simulation for solid-liquid phase change phenomenon of phase change material under constant heat flux. *Int. J. Heat Mass Transf.* **2015**, *86*, 197–206. [CrossRef]

20. Eshraghi, M.; Felicelli, S.D. An implicit lattice Boltzmann model for heat conduction with phase change. *Int. J. Heat Mass Transf.* **2012**, *55*, 2420–2428. [CrossRef]

21. Huang, R.; Wu, H.; Cheng, P. A new lattice Boltzmann model for solid-liquid phase change. *Int. J. Heat Mass Transf.* **2013**, *59*, 295–301. [CrossRef]

22. Sadeghi, R.; Shadloo, M.S. Three-dimensional numerical investigation of file boiling by the lattice Boltzmann method. *Numer. Heat Transf. Part A Appl.* **2017**, *71*, 560–574. [CrossRef]

23. Sadeghi, R.; Shadloo, M.S.; Jamalabadi, M.Y.A.; Karimipour, A. A three-dimensional lattice Boltzmann model for numerical investigation of bubble growth in pool boiling. *Int. J. Heat Mass Transf.* **2016**, *79*, 58–66. [CrossRef]

24. Chatterjee, D.; Chakraborty, S. An enthalpy-based lattice Boltzmann model for diffusion dominated solid–liquid phase transformation. *Phys. Lett. A* **2005**, *341*, 320–330. [CrossRef]

25. Li, D.; Tong, Z.X.; Ren, Q.; He, Y.L.; Tao, W.Q. Three–dimensional lattice Boltzmann models for solid–liquid phase change. *Int. J. Heat Mass Transf.* **2017**, *115*, 1334–1347. [CrossRef]

26. Latif, M.J. *Heat Conduction*, 2nd ed.; Springer: Berlin, Germany, 2009; ISBN 9783642012662.

27. Côté, J.; Konrad, J.M. A generalized thermal conductivity model for soils and construction materials. *Can. Geotech. J.* **2005**, *42*, 443–458. [CrossRef]

28. Chen, P.P.; Bai, B. SPH numerical simulation of moisture migration caused by temperature in unsaturated soils. *Eng. Mech.* **2016**. [CrossRef]

29. Shamsundar, N.; Sparrow, E.M. Analysis of multidimensional conduction phase change via the enthalpy Model. *ASME Trans. J. Heat Transf.* **1975**, *97*, 333–340. [CrossRef]

30. Guo, K. *Numerical Heat Transfer*; Anhui Science and Technology Publishing House: Hefei, China, 1987. (In Chinese)

31. Qian, Y.H.; D'Humières, D.; Lallemand, P. Lattice BGK model for Navier-Stokes equation. *Eur. Lett.* **2007**, *17*, 479. [CrossRef]

32. Wang, M.; Pan, N.; Wang, J.; Chen, S. Mesoscopic simulations of phase distribution effects on the effective thermal conductivity of microgranular porous media. *J. Colloid Interface Sci.* **2007**, *311*, 562–570. [CrossRef] [PubMed]

33. Zhaoli, G.; Chuguang, Z.; Baochang, S. Non-equilibrium extrapolation method for velocity and pressure boundary conditions in the lattice Boltzmann method. *Chin. Phys.* **2002**, *11*, 366–374. [CrossRef]

34. Qiu, F. Studies on Basic Theories and Technologies for Artificial Thawing of Artificial Frozen Soil. Master's Thesis, Tongji University, Shanghai, China, March 2011.

energies

MDPI

Review

Global CO$_2$ Emission-Related Geotechnical Engineering Hazards and the Mission for Sustainable Geotechnical Engineering

Ilhan Chang [1], Minhyeong Lee [2] and Gye-Chun Cho [2,*]

[1] School of Engineering and Information Technology (SEIT), University of New South Wales (UNSW), Canberra, ACT 2600, Australia
[2] Department of Civil and Environmental Engineering, Korea Advanced Institute of Science and Technology (KAIST), Daejeon 34141, Korea
* Correspondence: gyechun@kaist.edu; Tel.: +82-42-869-3622

Received: 10 June 2019; Accepted: 28 June 2019; Published: 3 July 2019

Abstract: Global warming and climate change caused by greenhouse gas (GHG) emissions have rapidly increased the occurrence of abnormal climate events, and both the scale and frequency of geotechnical engineering hazards (GEHs) accordingly. In response, geotechnical engineers have a responsibility to provide countermeasures to mitigate GEHs through various ground improvement techniques. Thus, this study provides a comprehensive review of the possible correlation between GHG emissions and GEHs using statistical data, a review of ground improvement methods that have been studied to reduce the carbon footprint of geotechnical engineering, and a discussion of the direction in which geotechnical engineering should proceed in the future.

Keywords: global warming; climate change; greenhouse gas; carbon dioxide; extreme precipitation; disaster; geotechnical engineering hazard; ground improvement; soil stabilization

---

1. Introduction

Greenhouse gas (GHG) emissions from human activities, especially carbon dioxide (CO$_2$) emissions from burning fossil fuels have continuously increased since the late 19th century and are strongly related to global economic growth and the population explosion [1]. Recent studies provide strong evidence of the progressive climate change brought about by the anthropogenic increase in GHG emissions [1]. GHGs in the Earth's atmosphere play a major role in temperature control by absorbing approximately 20% of the radiant heat emitted from the Earth's surface and then releasing it back to the surface [2,3]. Greenhouse gas emissions, and the accompanying atmospheric concentrations of CO$_2$, have continuously increased (Figure 1) [4,5]. The atmospheric concentration of CO$_2$, which was 280 ppm in 1750, has shown a 42% increase to 400 ppm in 2015 [1,3,4]. As a result, global mean temperatures are continuously rising (Figure 2) where 2015, 2016, and 2017 were recorded as the most warmest years since 1880 [5–8].

In response and as part of global efforts to reduce GHG emissions (particularly CO$_2$), 197 countries represented in the United Nations Framework Convention on Climate Change adopted the Paris Agreement in 2015 and committed to cutting emissions, with the aim of maintaining global mean temperatures below 2 °C [9]. However, no noticeable reduction or effort has yet been made, and average CO$_2$ emissions are expected to continue to rise as a result of industrial growth in developing countries and the global urbanization trend [10]. Thus, more effective action is required to maintain global mean temperatures below 2 °C [11].

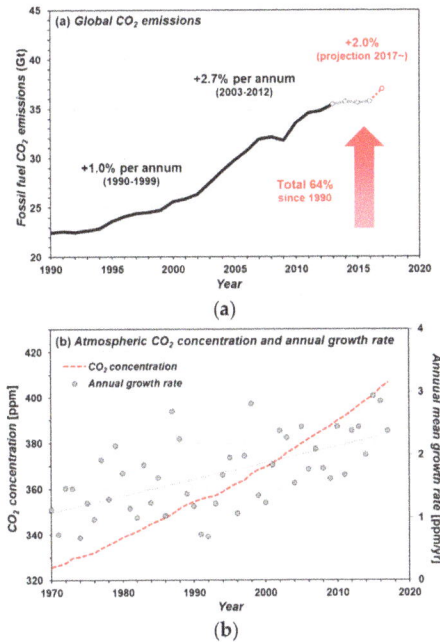

Figure 1. Global CO_2 emission status: (**a**) Trend of global CO_2 emissions from fossil fuels. Reproduced with permission from Le Quéré et al., Global Carbon Budget 2017 in Earth System Science Data; published by Copernicus Publications, 2018 [5]; (**b**) Atmospheric CO_2 concentration and annual growth. Reproduced with permission from Ed Dlugokencky and Pieter Tans, Trends in atmospheric carbon dioxide (www.esrl.noaa.gov/gmd/ccgg/trends/); published by the National Oceanic & Atmospheric Administration (NOAA) [4].

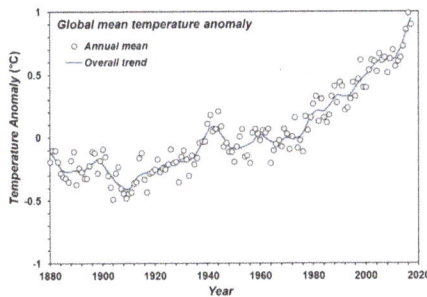

Figure 2. Global mean temperature pattern from 1880 to 2017. Reproduced with permission from the Jet Propulsion Laboratory (JPL), Global land-ocean temperature index (https://climate.nasa.gov/vital-signs/global-temperature); published by the National Aeronautics and Space Administration (NASA) [8].

The increase in global atmospheric CO_2 concentration and corresponding mean temperature of the earth alters global water circulation, which is followed by unexpected weather events (e.g., heavy downpour, drought) and a rise in sea levels [1,12]. In other words, the pattern of hydrologic climate events (e.g., frequent localized heavy rain and intensive storms) is changing and induces unsuspected geotechnical engineering hazard (GEH) events (i.e., landslides, ground subsidence, levee failures, soil degradation, and coastal erosion) [13]. Thus, since the end of the 20th century, the occurrence of, and damage from, GEH events around the world has rapidly increased, along with huge social and economic losses [14–17]. For instance, in the United States, 219 natural disasters with damage exceeding $1.5 trillion occurred in 2017 alone (Figure 3a) [18,19]. The economic damage of disasters has also drastically increased in South Korea since 1990 and has become an important national issue of safety (Figure 3b) [20,21]. In response, more countries require geotechnical engineering implementation for damage recovery or disaster mitigation [22].

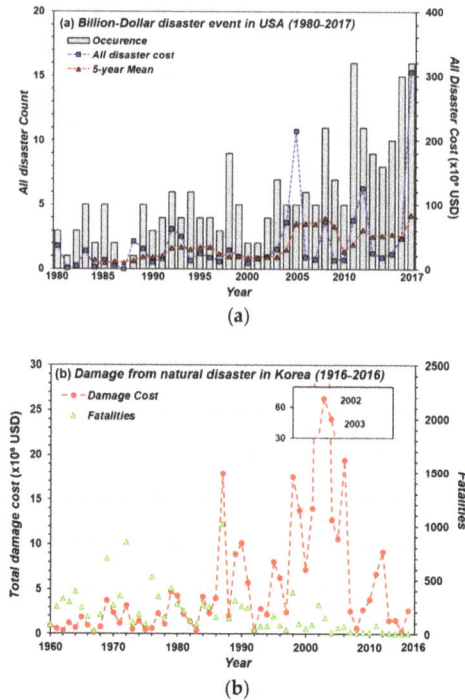

(a)

(b)

Figure 3. Disaster occurrences, including flooding, severe storms, drought, tropical cyclones, winter storms, freezing, and wildfire, with relevant damage costs: (**a**) The U.S. billion-dollar disaster record (1980–2017). Reproduced with permission from the National Oceanic and Atmospheric Administration (NOAA) of the United States, U.S. Billion-Dollar weather and climate disasters (https://www.ncdc.noaa.gov/billions/time-series); published by National Centers for Environmental Information (NCEI) [19]; (**b**) Total disaster damage cost and fatalities in South Korea (1916–2016). Reproduced with permissions from 1) Ministry of the Interior and Safety, Annual Disaster Report 2016; published by the Ministry of the Interior and Safety of the Republic of Korea Government, 2017 [20], and 2) National Emergency Management Agency, Annual Disaster Report 2009; published by National Emergency Management Agency of the Republic of Korea Government, 2010 [21].

Therefore, a comprehensive understanding of GEHs in the context of climate change and CO_2 emissions is required. Furthermore, sustainable ground improvement methods that can respond to GEHs while reducing the CO_2 footprint should be introduced and implemented in geotechnical engineering [23]. This study aims to provide an overview on the effect of climate events on GEHs and a statistical review of the correlation between the occurrence of, and damage from, GEHs and CO_2, based on historic disaster data. The status and challenge of several ground improvement methods to replace high CO_2 emitting soil binders (e.g., cement) are also summarized, and the necessity of an environmentally friendly perspective in geotechnical engineering is addressed.

2. Relationship between Climate Change and Geotechnical Engineering Hazards

2.1. Climate Change Issues Related to Global Warming

Two major climatic issues are associated with global warming, which is accelerated by additional GHGs (Figure 4): (1) Extreme precipitation and (2) sea level rise [17,24].

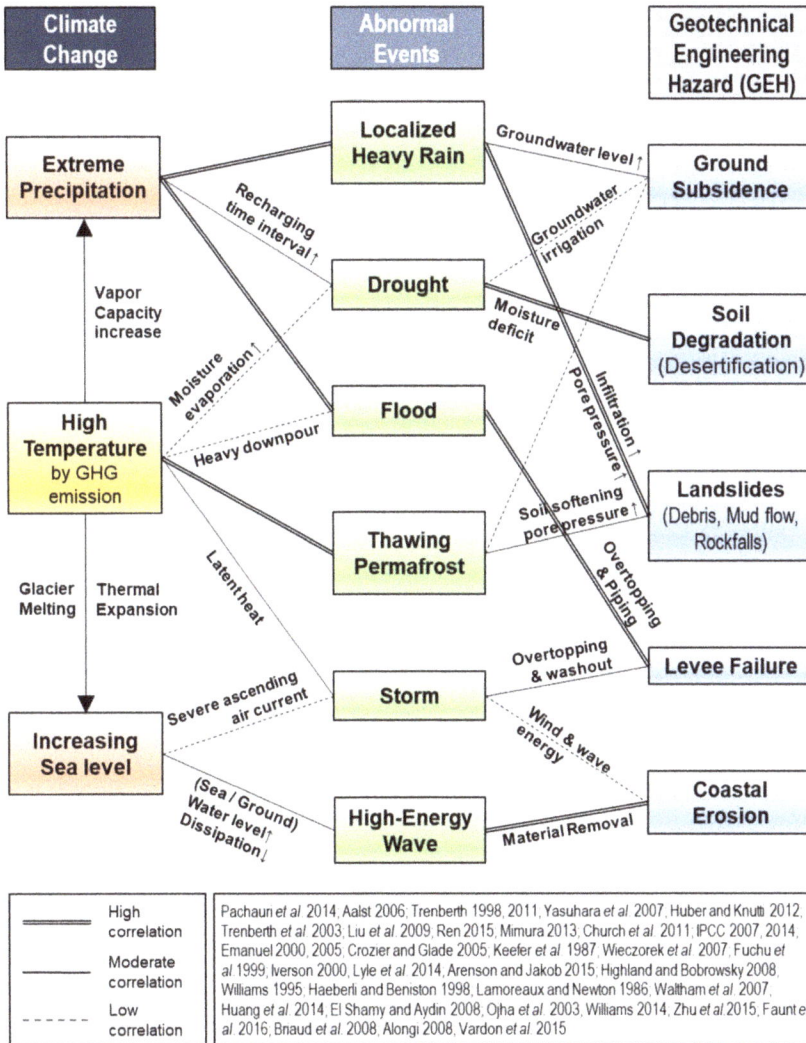

Figure 4. Geotechnical engineering hazard (GEH) events triggered by climate change [1,12,13,17,24–53].

Extreme precipitation takes place when warmer temperatures allow the atmosphere to hold more water vapor. The atmosphere is able to contain more water vapor because its capacity increases by 7% when the atmospheric temperature rises by 1 °C [13,25–27]. As more water evaporates into the atmosphere, clouds with heavy concentrations of water vapor can render localized and heavier downpours, while other places experience drought. Moreover, the intervals between wet periods in the water circulation process can be disturbed and generate more extreme precipitation events, as illustrated in Figure 5 [28]. Localized heavy rainfall and droughts generated by extreme precipitation can cause landslides, ground subsidence, and soil degradation. For example, successive and torrential heavy downpours in southwestern Japan in June and July of 2018 triggered landslides, mudslides, and flash flooding, causing 225 deaths [54,55].

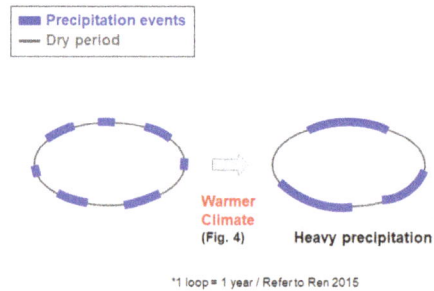

Figure 5. Illustration of extreme precipitation pattern in a warmer climate, reproduced with permission from Ren Diandong, Strom-triggered Landslides in Warmer Climates; published by Springer, 2015 [28]. Blue circles indicate precipitation events (wet days), and gray lines indicate the dry periods in between.

Meanwhile, sea level rise is mainly caused by the thermal expansion of sea water and the melting of glaciers. Latent heat is transferred from the atmosphere to the ocean as the atmospheric temperature becomes warmer due to high GHG concentrations. This increased heat capacity of the ocean, in combination with an inrush from the melting of mountain glaciers and ice sheets in Antarctica and Greenland [29], raises the average sea water level, which causes the sea level to rise [29–31]. As a result, an increased sea level creates more severe ascending air currents, helping form intensive, larger, and longer-lasting storms with heavy rains, such as hurricane Florence, which delivered nearly three feet of rain on North Carolina in 2018, causing severe damage [34,56,57]. In addition, an increased sea level strengthens waves reaching the shore by reducing the wave energy dissipation related to friction, which depends on the depth of the coastal floor [17]. The combination of high-energy storms and waves generated by sea level rise leads to severe and simultaneous erosion and flooding events that commonly cause levee failure, coastal decomposition, and ground subsidence [58].

2.2. Effect of Abnormal Climate Events on Ground Properties and Geotechnical Engineering Hazards

The geotechnical effect of several abnormal climate events on ground properties, and the GEHs resulting from them, are summarized in Table 1. In detail, water from localized heavy rain penetrating the ground increases the excess pore pressure of the ground in a short time (the soil suction value decreases). This reduces the effective stress between soil particles, thereby weakening the shear strength of the inclined ground and (in combination with the overburden caused by trapped rain in the active layer of the ground) leads to landslides or slope failures [32,35–38]. The reduction in ice-cementation bonds between soil particles, due to thawing in permafrost sediments as the global atmosphere gets warmer, can also assist in wet mass movement, such as debris flow [33,39–41]. Likewise, the degradation of permafrost in fractured rock mass creates concerns about rockfalls, due to long-term changes in stress distribution caused by reduced strength and increased permeability in rock masses [42].

Table 1. Overview of abnormal climate events and their effect on ground properties.

Abnormal Climate Event	Effect on Ground Properties	Related Geo-Hazard	References
Extreme Precipitation — Localized Heavy Rain	Pore pressure increase → Soil suction value decrease → Soil effective stress and shear strength reduction	Landslide	
	Higher infiltration into surface layer → Unit weight increase above potential failure surface → Increased driving force inducing downward movement	Landslide	[35–38]
	Flood → Rise of seepage line or overtopping, which increases the degree of saturation due to infiltration → Pore pressure increase → Void ratio and hydraulic conductivity increase → Effective stress decrease	Levee failure (breach, piping)	[45–47,59]
	Extreme groundwater table variation and dissolution of soluble geomaterials (e.g., $CaCO_3$)	Ground Subsidence (including sinkholes)	[43,44]
Drought	Severe evaporation → Moisture deficit in surface soil → External soil shrinkage and internal erosion → Vegetation cover decay and soil vulnerability (erosion) increase	Soil degradation (Desertification) Ground subsidence	[48–50,53]
Thawing Permafrost	Destruction of ice-cementation bonds and unfrozen water increase in soil → Shear strength decrease	Landslide Heaving and subsidence (including thermokarst)	
High Average Temperature	Higher water level on coasts → Less wave energy dissipation → Higher wave energy approaching coasts → Air trapping in pore spaces and compression by waves → Weakening of soil particle interaction → Break off and coastal erosion increase	Coastal Erosion Coastal Landslide	[33,39,41,42]
Sea Level Rise	Latent heat energy and vapor transfer to the air → Severe and higher air ascending stream (heavy storm) → Extensive and frequent inundation by storm surges in coastal regions → Overtopping and washing out → Erosion and failure	Coastal Erosion Levee Failure	[17,51,52]

However, downpours in certain areas can lead to a dramatic change in the groundwater level. For instance, in karst terrain, the limestone geologic compositions can be easily dissolved by water, and calcium carbonate ($CaCO_3$) dissolution can accelerate sudden ground subsidence events, such as sinkholes. These events can also occur in urban areas. For example, a groundwater change can lead to the loss of soil near water pipelines with leaks [43,44].

Rainfall intensity increase due to changes in precipitation patterns can lead to frequent flooding and other GEHs, such as failures of levees and erosion in riverine areas. In drastic riverine flooding caused by heavy rain, the water level exceeds the allowable design capacity of levees or embankments, which generally results in overtopping, whereby overflow water erodes the end of a slope, leading to failure [45]. In addition, internal erosion in a levee and an excessive flow rate of water with a tractive force eroding away the bottom and lateral surfaces can contribute to severe earthen levee failures [46,47].

Meanwhile, drought caused by an extended dry period with limited precipitation results in a moisture deficit in the surface layer of soil. Surface desiccation and soil shrinkage decreasing surface vegetation cover make the land more vulnerable to soil erosion, leading to soil degradation and desertification [48,53]. Some recent studies have also reported that irregular groundwater irrigation due to water shortages can render ground subsidence [49,50].

In coastal areas, unprecedented strong storms due to sea level rise and warmer temperatures can weaken the ground strength and escalate surface erosion [51]. In particular, high-energy waves accelerate coastal erosion and can also result in damage to shore structures through a decrease in load-bearing capacity [17]. In addition, sea level rise and warming sea waters are speeding up coastal erosion by destroying coastal ecosystems (e.g., mangroves and reefs) that attenuate waves and prevent the washing away of particles in coastal areas [52].

3. Statistical Trends of CO_2 (Climate Change) Emission and Geotechnical Engineering Hazards

3.1. Status of Geotechnical Engineering Hazards

The Emergency Events Database (EM-DAT) provided by the Centre for Research on the Epidemiology of Disasters (CFRD) is among several widely used international disaster databases. EM-DAT was created in 1988 with support from the World Health Organization and the Belgian government, which provides overall disaster data from United Nations agencies, U.S. government agencies, research centers, and the press every year [60]. As EM-DAT is publicly available and has been used in a number of scientific studies, it is used in this study for statistical review [61–64]. The database in EM-DAT is mainly classified by biological, geophysical, climatic, hydrologic, meteorological, and extraterrestrial groups, including several subgroups in each main group [60]. This study focused on the data for landslides (wet mass movements), floods, wave action, and droughts (excluding dry mass movement or ground subsidence by earthquake) to analyze the statistics of GEHs from the perspective of climate change and geotechnical engineering.

The EM-DAT database provides global disaster data from 1900 to the present. However, the old data may not be reliable, due to the inadequate standardization of the data collection and analyzing methods up to the middle of the 20th century. Thus, the authors decided to focus on the occurrence and damage data of GEHs since the 1960s.

3.2. Relationship between CO_2 Emissions and Geotechnical Engineering Hazards

The damage scale of disasters may be attributed to multiple factors, including anthropogenic influences on global warming and climate change, as well as socioeconomic conditions, such as infrastructure development level and readiness of national or local disaster confrontation systems [65]. Still, there is no doubt that climate change strongly correlates to the frequency increase of severe GEH events, and the simultaneous global population growth during the past century must be considered when interpreting damage scale and socioeconomic impacts (e.g., damage cost, casualties, and affected populations) of GEHs [66].

Figure 6a shows the overall incidence, damage, and affected populations of global GEHs from 1900 to 2017 [67]. All indices in Figure 6a show significant increases since the 1960s. Meanwhile, damage scales (cost and affected people) adjusted for the global population in each year (in cost or people per million people) are plotted in Figure 6b, demonstrating climate change's effects on the significant rise in GEH-related damage. The occurrence of GEHs among different continents is shown in Figure 7. All continents show a simultaneous and continuous increase in GEHs occurrences since the 1960s. In particular, occurrences in Asia grew most rapidly among the continents, indicating that the monsoon region is affected by frequent floods and landslides caused by recent climate change [68]. If the frequency and damage of GEHs were similar to levels prior to the damage, indices may have either been reduced or remained steady due to the social economic growth and technology development [69–72]. However, as most disasters are unpredictable, the positive increase in the data indices could mean that the magnitude and intensity of each unforeseen GEH has become much stronger. In other words, these results show that abnormal climate phenomena of severe magnitudes have increased, and that climate change is a direct cause of GEHs.

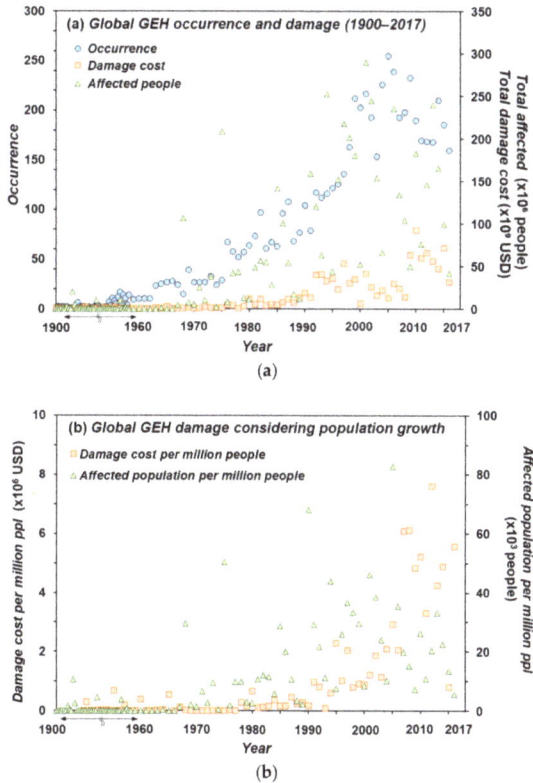

Figure 6. Global geotechnical engineering hazard status (1900–2017): (**a**) Occurrence and total damage indices; (**b**) Damage cost and affected population per million people. Reproduced with permission from the Centre for Research on the Epidemiology of Disasters (CRED), Emergency Events Database (EM-DAT); published by the Université catholique de Louvain (UCL) [67].

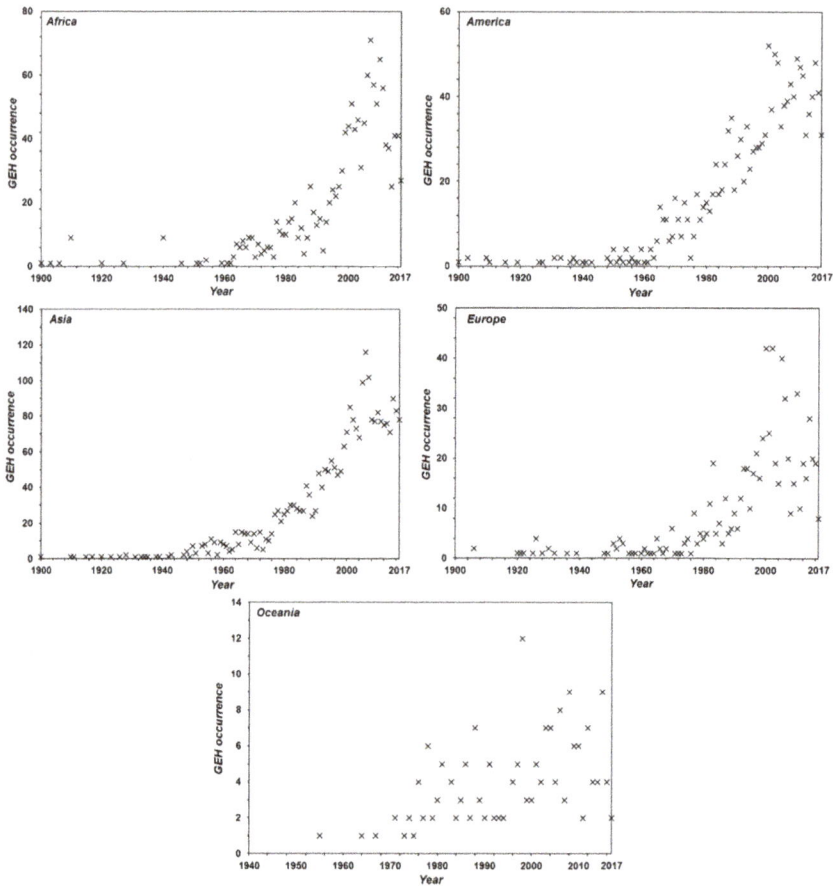

Figure 7. Geotechnical engineering hazard occurrence by continent from 1900–2017. Reproduced with permission from the Centre for Research on the Epidemiology of Disasters (CRED), Emergency Events Database (EM-DAT); published by the Université catholique de Louvain (UCL) [67].

As mentioned above, CO_2 emissions from human activity, especially the burning of fossil fuels, are one of the main contributors to global warming and climate change [1]. The recent rapid increase of atmospheric CO_2 concentration—324 ppm in 1970 to 406 ppm in 2017, representing a 25% rise—is known to induce climatic events of greater abnormality and severity [73], in line with data scattering associated with higher CO_2 concentration in Figure 8. Most emitted GHG, including CO_2, exist in the atmosphere for several decades [3]. Furthermore, since global CO_2 emissions have been continuously increasing and are unlikely to be flat in the near term, the climate change phenomenon that has been recently observed is regarded as the beginning [1] Therefore, the reduction of CO_2 emissions is an essential way to mitigate GEH damage to human civilization, from a geotechnical engineering aspect.

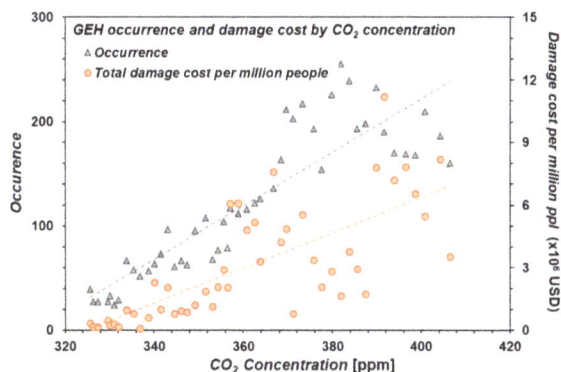

Figure 8. Correlation between CO_2 concentration and geotechnical engineering hazards, in terms of occurrence and total damage per million people. Reproduced with permission from the Centre for Research on the Epidemiology of Disasters (CRED), Emergency Events Database (EM-DAT); published by the Université catholique de Louvain (UCL) [67] and permission from Ed Dlugokencky and Pieter Tans, Trends in atmospheric carbon dioxide (www.esrl.noaa.gov/gmd/ccgg/trends/); published by the National Oceanic & Atmospheric Administration (NOAA) [4].

4. The Response to Geotechnical Engineering Hazards and the Necessity of an Environmentally Friendly Method

4.1. Contribution from Geotechnical Engineering to Reduction of CO_2 in the Earth

Over the last three decades, much geotechnical engineering research has been conducted to mitigate CO_2 emissions from fossil fuels [74].

One representative approach to directly reduce already-emitted atmospheric CO_2 is carbon capture and storage (CCS), including geological CO_2 storage (GCS). GCS aims to inject captured atmospheric CO_2 into underground geological media, such as oil and gas fields, coal layers, deep saline aquifers, and hydrate bearing sediments [74,75]. Compared to the other CCS techniques, GCS has the advantage of large capacity and additional merit in enhancing oil recovery. However, GCS poses challenges, including high cost, long-term leakage, and the possibility of rendering subsea GEHs [75].

CCS and GCS technologies are the predominant tactics in reducing the present atmospheric CO_2, with less effect on mitigating GEHs triggered by CO_2 emission-related climate change. Thus, this section will focus on current attempts in ground improvement by not using high CO_2 emitting cement in geotechnical engineering practices.

4.2. Ground Improvement and CO_2 Emissions Related to Cement

Since most GEHs are related to soil strength reduction due to changes associated with water, geotechnical engineers have been studying various methods of ground improvement, to increase the strength of the ground. For instance, retaining walls [76], geosynthetic products [77], and anchors with grout (called soil nailing) [78,79] are installed to increase the stability of slopes. Also, the strength of soft, clayey soils has been improved through electrokinetic stabilization with chemical grouting to prevent slope failures (e.g., landslides) [80].

To prevent levee failures, concrete pilings or geosynthetic products are constructed to strengthen the levee structure and resist against overtopping or internal erosion [81]. Ground subsidence, which is mainly affected by changes in the groundwater level, can be mitigated via cement or lime based deep mixing or grouting practices [82], while chemical binders (including cement and polyurethane) are commonly used to prevent erosion and cliff failure in coastal regions [83]. Thus, it should be noted that cement and chemical binders are mostly used for geotechnical ground improvement, to respond to GEHs. Cement has various advantages in terms strengthening, durability, and economic aspects, thus it holds a dominant position among other construction materials in civil and construction engineering practices. However, questions have recently been posed about the long-term environmental impact of cement, despite its many advantages.

Cement production emits CO_2 through two main processes: the kiln calcination ($CaCO_3$ + heat $\rightarrow$ $CaO + CO_2$) and the combustion of fossil fuels for heating. Generally, about one ton of CO_2 is generated to produce a single ton of cement [84]. According to data from the U.S. Geological Survey, 4.2 gigatons of cement are produced worldwide per annum, and the percentage of cement-related CO_2 emissions in total CO_2 emissions has reached almost 10%, more than doubling from 4% in 1970 (Figure 9) [85]. In geotechnical engineering practices, ground improvement processes such as mixing and grouting are reported to contribute about 0.2% of entire global CO_2 emissions [23]. Although global efforts to reduce CO_2 emissions were initiated after The Paris Agreement in 2015, cement production is expected to grow, due to the huge demand for traditional civil engineering materials, particularly in China, India, and large parts of the developing world, given the global urbanization trend [84]. In geotechnical engineering perspectives, ironically, cement that releases CO_2 in its production is used to prevent and recover from GEHs related to climate change caused by CO_2. Moreover, other environmental problems, including alkalization of the soil (affecting ecosystems, urban runoff, and vegetation levels), demonstrate the need for environmentally friendly and sustainable alternatives to cement to reduce the CO_2 footprint. In response, various geotechnical approaches for alternatives to cement, such as chemical mixtures, geopolymers, geosynthetics, microbial organisms, and biopolymers, have recently been investigated.

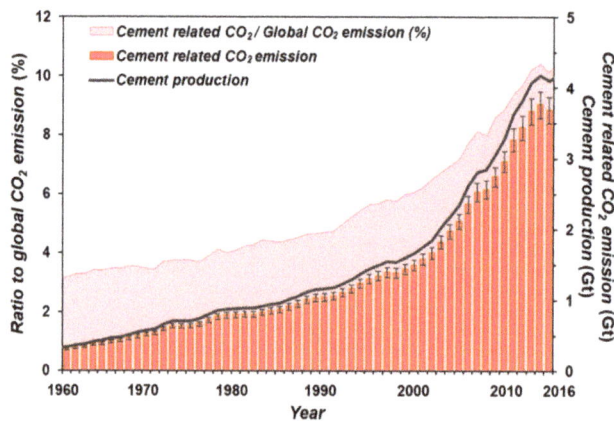

Figure 9. CO_2 emissions related to the production of cement and its ratio to total CO_2 emissions. Reproduced with permission from Thomas D. Kelly and Grecia R. Matos, Historical statistics for mineral and material commodities in the U.S.; published by the United States Geological Survey (USGS), 2015 [85].

4.3. Recent Research on Environmentally Friendly Ground Treatment Methods

Various geotechnical approaches for replacing cement are listed in Table 2.

Table 2. State-of-the-art attempts to reduce cement usage in geotechnical engineering.

Properties	Chemical Stabilizer	Geopolymer	Geosynthetics	Microbiologically Induced Calcite Precipitation	Biopolymer
Methodology	Injection or spraying and mixing before compaction; Chemically synthesized polymers	Mixing, injecting, or spraying of alkali activated pozzolans; Alumina-silicate (i.e., pozzolanic materials) and alkali or alkali earth substance	In situ installation of synthetic materials; Synthetized polymer products	Injecting bacteria and nutrient solution into the ground; Microbial (bacterial) and urease enzyme	Direct mixing, injecting, or spraying of biopolymers; Dry (powder type) of hydrogel (solution) biopolymers
Materials and Mechanism	(i.e., acrylamide-based anionic polyelectrolytes); Ionic bonding with soil particles or interparticle cementation	Alkali silicate activation (polycondensation)	Tensile strength enhancement and fluid flow control in soil	Biologically driven $CaCO_3$ precipitation (cementation, pore-clogging)	Particle aggregation of inter-particle bonding through hydrogen and ionic bonding
Geotechnical Effects	-Strength improvement and density increase; -Reduced sensitivity to water (plasticity)	-Void ratio reduction by geopolymerized gel filling and increased bulk density	-Separation, filtration, and drainage of water in soil; -Tensile strengthening; -Impeding flow of liquid or gas	-Improvement in soil matrix stiffness and initial shear strength; -Hydraulic conductivity control	-Cohesion and strength increase (biopolymer–soil matrix formation); -Permeability reduction
Advantages	-Prevention of detachment by erosion and runoff; -Encouraged seed germination; -Flocculants for wastewater treatment; -Increased sweep efficiency in oil recovery	-Lower CO_2 emissions than cement; -Resistance to acid, sulfate, and freeze-thaw attack; -Usage of industrial by-products (fly and bottom ashes)	-High durability; -Easy transportation and site installation; -High tensile strength, flexibility, and imperviousness; -Various ranges of applications	-Low energy consumption, with a low carbon footprint; -Flexible implementation in soil due to easy control of the treatment process, using bacteria; -Chemical characteristic of soil grains do not alter	-Low carbon footprint and biodegradability; -Low binder quantity; -Sufficient quality control; -Erosion reduction and vegetation improvement
Limitations and Challenges	-Contamination concerns into soil and ground water; -High material cost; -Infeasible for deep/thick ground treatment	-Lack of standards for tests and production; -Lack of geotechnical applications; -Needs heat process (about 60 °C) in the field	-Material-dependent strength; -Non-biodegradable; -Inappropriate for significant depths in the ground	-Inappropriate for fine soils; -Consistent quality control; -Weakness against low pH; -Ammonia as a byproduct; -Few field applications	-Low economic feasibility; -High sensitivity to water; -Severe hydrogel swelling; -Concerns on long-term durability
Related Recent Research	-Monitoring of long-term effectiveness by measuring metal bioavailability and soil quality improvement; -Biomass silica stabilizer from agricultural waste; -Calcium carbide residue from acetylene production	-Attempt to use lime sludge from paper industry waste for paving blocks; -Soft marine clay stabilization by fly ash and calcium carbide residue-based geopolymer	-Nano clay combined geotextile for removing heavy metal or toxic manners; -Hybrid combined geosynthetics; -Sensor-embedded geosynthetics	-Field-scale test focused on surface applications for erosion and dust control; -Use of seawater as a calcium source (feasibility for marine applications)	-Casein from dairy waste as a new binder; -Inter-particle interaction characterization using microscopic devices; -Strength enhancement in wet conditions using crosslinking; -Economic feasibility improve
Reference	[86–92]	[93–101]	[102–108]	[109–117]	[23,118–127]

4.3.1. Chemical Stabilizers

After the 1950s, research commenced on nontraditional soil stabilization additives consisting of multiple chemical agents as a means of replacing traditional binding materials (i.e., cement and lime) in geotechnical engineering. One of these agents was lignosulfonate, a chemical stabilizer containing Na-, Ca-, and NH_3-lignosulfonate, which are synthetic materials from lignin used as cellulose fibers. When this material meets soil, it coats the soil particles with a thin adhesive film and bonds them together. In addition to its primary cementing effect, lignosulfonate forms ionic bonds with clay particles having electrically charged surfaces rendering a strengthening effect [86]. In other words, soil strength is enhanced and shrinkage–swelling is reduced. This approach is particularly effective in coarse and granular soils [87].

Other types of chemical stabilizers are anionic acrylamide-based polyelectrolytes, and the most common type is polyacrylamide (PAM). PAM establishes an effective interaction among soil particles through charge neutralization, bridging, and adsorption, using negative charges. Thus, PAM has been applied to multiple geotechnical engineering practices, such as flocculants, shale stabilizers, and thickening and binding agents. PAMs are especially used to reduce runoff, erosion, and soil sealing [88].

Salt stabilizers, including calcium and magnesium chloride compounds, and other polymer stabilizers, such as vinyl acetates, have been shown to improve strength, stabilize volume, and enhance waterproofing with sandy or clayey soils [86].

Chemical stabilizers are generally implemented by injecting them into or spraying them on the soil and mixing before compaction. However, they have the potential to contaminate the surrounding geo-environment and nearby groundwater, resulting in limited usage near drinking water sources, despite their merits [92]. Therefore, although recent studies have been conducted on applying sustainable and nontoxic chemicals, such as biomass silica and calcium carbide [89,90], to soil, the economic limitations of applying deep injections to a large-scale site, and the necessity of establishing criteria for the laboratory scale to predict effective in situ performance, should be further researched [91].

4.3.2. Geosynthetics

Geosynthetics are synthetized polymeric products used with soil, rock, or other earth material to solve geotechnical engineering problems [103]. They are generally classified into eight main product types: Geotextile, geogrid, geomembrane, geocomposite, geosynthetic clay liner, geonet, geofoam, and geocells [23]. Geosynthetics are generally prefabricated and transported to a site in the form of a roll package and installed directly in the ground. The biggest advantage of geosynthetics is their multiple functions. In detail, geosynthetics are installed in the transition zone of intermixed ground to serve a separation function, prevent intrusion between aggregates with different sizes, or increase the stability of soil–geosynthetics composites by applying tensile strength. They are also used to provide filtration and drainage by adjusting the fluid flow path, to address soil erosion and other geosynthetics (i.e., geomembrane) that are relatively weak against external damage, through stress relief [102]. The production of geosynthetics grew after 1970 due to their advantages of easy installation, transport, and handling, and because of their ability to reduce construction time and cost [102,104]. Also, some geosynthetics (e.g., polypropylene, polyester, polyethylene, and polyamide) are highly applicable to various geotechnical problems and have excellent biological and chemical resistance. Therefore, the geosynthetics market has steadily grown, and in 2017 the demand for these materials was approximately 5200 mm^2 worldwide [76].

Recently, studies have been conducted on multipurpose hybrid geosynthetics that combine two or more types of geosynthetics to improve their applicability. Sensor-embedded geosynthetics have also been investigated for long-term site monitoring [107,108]. However, geosynthetics have some limitations and challenges. The strength of soil geosynthetics mainly relies on that of the synthetized material itself, and they are applicable at shallow depths but are not appropriate at significant depths [23]. Although their durability is as good as a plastic material, their sustainability needs to be further discussed and verified, due to their ecotoxic effect on the environment from the leakage of additives and residual product from degradation of polymeric or metallic materials into the ground [104–106].

4.3.3. Geopolymers

Geopolymers, a substitute for Portland cement, are alkali-activated, cementitious binder materials produced by a reaction called geopolymerization that occurs between aluminosilicate materials containing high levels of silicon and aluminum oxides (i.e., slag, fly ash, and metakaolin) and alkali-activating agents (i.e., alkali hydroxides) [93,98]. Alkali-activated geopolymers consist of three-dimensional structures of sodium aluminosilicate hydrate (N-A-S-H) gel along with calcium-silicate-hydrate (C-S-H) gel, which make up the Portland cement [94]. These geopolymerized gel binders, formed through a polycondensation process, fill the pores between soil particles to reduce the void ratio and increase bulk density, resulting in enhanced strength. Geopolymers have an environmental advantage because industrial wastes, such as fly ash or blast furnace slag, are used as the raw materials [95] and because the consumption of heat energy is smaller than conventional Portland cement, thereby reducing CO_2 emissions [99].

According to recent studies, however, one of the typical geopolymers based on fly ash has lower initial strength characteristics compared to Portland cement, due to its slow and time-dependent strength gain rate [97,101], so heat-curing treatments to increase strength lead to in situ geotechnical limitations [96]. Despite a numbers of studies having been conducted to verify the strengthening efficiency of geopolymers compared to cement-based methodologies, further research about standards for testing and production, including the generalization of the water/geopolymer ratio, the Si/Al and Na/Al ratios, and the bond between reinforcement and geopolymer paste, are required for reliable in situ applications [100].

Although ground improvement using geopolymerized binders and chemical additives, through jet grouting or spraying, aid in enhancing shear strength, their lack of feasibility in large-scale applications, due to potential contamination of the surrounding environment (including the effect on the pH of soil and groundwater) [109], has led to the development of new, alternative approaches. For example, biological soil stabilization methods relying on microbial-induced calcite precipitation (MICP) and biopolymers have recently been developed to achieve environmental sustainability.

4.3.4. Microbiologically Induced Calcite Precipitation

Microbiologically induced calcite precipitation (MICP) is one of the most recognized biological ground treatments and uses biologically induced $CaCO_3$ precipitation with urea hydrolysis through the metabolism of bacteria such as *Bacillus pasteurii* and *Sporosarcina pasteurii* [23]. Carbonate crystals produced by urea hydrolysis and precipitated near particles act as cementitious inter-particle bonding agents, strengthening the stiffness and shear strength of the soil matrix by blocking the pores and, furthermore, reducing hydraulic conductivity. Through these mechanisms, called bio-cementation and bio-clogging, MICP can be flexibly implemented to solve geotechnical problems [113]. Also, MICP involves less energy consumption, which results in a low carbon footprint compared to traditional ground improvement methods [113]. Recent studies have used an enzymatic reaction with calcium chloride instead of a direct injection of grown bacteria in an attempt to enhance the production rate of carbonate [111,112]. Other research has involved the removal of heavy metals [114], surficial application of MICP for erosion and dust control [115], and usage of seawater as a nutrient source to attain higher carbonate precipitation [116]. However, most MICP research so far has been conducted on the laboratory scale using coarse-grained soils in which the pores are relatively large, and shows applicability limitation for clayey soils [110]. Also, the non-uniform distribution of precipitated $CaCO_3$ in field conditions and the emission of ammonia as an end product are future challenges for MICP. The relatively high price of bacterial nutrients is another issue. Therefore, further research should be conducted to resolve these limitations and to scale up to in situ and industrial-level production studies [109,117].

4.3.5. Biopolymers

Another biological soil treatment method involves excretory products from living organisms, called biopolymers, which have recently begun to gain attention for their soil-stabilization potential in the geotechnical engineering field [23]. Biopolymers, which are widely used in the food and pharmaceutical industries, are defined as an assembly of monopolymers produced from biological organisms. Polysaccharides, such as cellulose, starch, Xanthan, β-glucan, and gellan gum, are among several types of biopolymers that have been recently examined in the geotechnical engineering field with the purpose of soil strengthening and hydraulic conductivity control [119–121,128]. In previous studies, biopolymers increased the compressive and shear strength of soil through hydrogen-based chemical bonding between the biopolymers and soil particles. Bonding by electrostatic attraction between biopolymers and clay particles generates a noticeable strengthening effect [121,124]. For instance, a gellan gum and soil mixture (kaolinite clay) showed greater strength than a 10% cement mixture, in spite of the small content of biopolymers (under 2%) [121]. Furthermore, biopolymers are hydrophilic, and they absorb water to form a water zone in wet conditions, resulting in a reduction in permeability by filling the pores in soil with an expanded biopolymer gel [124]. Therefore, biopolymers have the potential to prevent desertification by increasing soil strength, reducing particle erosion, and assisting plant growth with their high water-retention capabilities [123].

From an environmental perspective, biopolymers produce few CO_2 emissions and can be naturally decomposed (biodegradable) with no harmful effect on the geo-environment or groundwater to which they are applied, so biopolymers are a promising option as environmentally favorable and sustainable materials for the future [23]. The biggest distinction from traditional Portland cement and the other alternatives to cement is that biopolymers can provide higher strength for smaller binder content [23]. According to previous research, the strength of biopolymer-treated soil with a biopolymer to soil content in mass of ~0.5–1% can be similar to, or stronger than, Portland cement and geopolymer cement, which must make up at least 10% of the content in soil-improvement applications [120]. Also, unlike MICP, biopolymers can be externally cultivated, enabling efficient mass production with high quality. They can also be applied using multiple methods, including injection, spraying, and mixing, which means they have the potential to be applied in various geotechnical field applications (e.g., for deep mixing, slope reinforcement, quick blockage of water inflow, and vegetation improvement). However, biopolymers are less economically feasible, due to their expensive global market price compared to cement. They are being used in other industries (e.g., food, pharmaceuticals, and cosmetics) that require pure and good-quality biopolymers, but they are not common in the geotechnical engineering field, so production costs are high. According to a recent survey, the unit costs of some biopolymers are declining with the growth of the biopolymer market [122], and biopolymers that have a relatively rough quality may be suitable for geotechnical engineering, so they are likely to be more competitive in the future [129]. However, the challenges are that not enough research has been conducted to provide specific guidelines for field applications and that biopolymers have a durability problem, due to strength reduction in wet conditions, that needs to be resolved [125]. In response to these issues, recent research has attempted to maintain soil strength by using thermo-gelation biopolymers [121,125], water-familiar, protein-based biopolymers (i.e., casein) [126], and cross-linking biopolymers [130].

5. Conclusions

This study provides a statistical review of the increase in GEHs resulting from CO_2 emissions from a geotechnical engineering perspective. Global warming, accelerated by increases in CO_2 emissions, creates abnormal climate events around the world that change the properties of soil over the short and long term, resulting in GEHs such as landslides, ground subsidence, levee failures, soil degradation, and coastal erosion. The occurrence and damage costs of GEHs have increased because of climate change, and these patterns are positively correlated with the increase in atmospheric CO_2 concentration. Meanwhile, cement has been the most widely used material as a ground improvement method in geotechnical engineering in response to GEHs. Cement accounts for 10% of global CO_2 emissions, so the geotechnical engineering field in the 21st century faces the ironic situation of simultaneously using CO_2-emitting materials to prevent the increase of GEHs related to CO_2 emission increase. Therefore, ground improvement methods implementing alternatives to cement, such as chemical additives, geosynthetics, geopolymers, MICP, and biopolymers, to reduce the global carbon footprint have been examined for sustainable geotechnical engineering. These attempts to find suitable alternatives are not yet fully satisfying, in terms of strength, economic feasibility, and field applications, and each method has its own limitations and challenges. Consequently, geotechnical engineering research should be more actively advanced in the direction of environmentally friendly and sustainable geotechnical materials that can contribute to reducing CO_2 emissions and relevant threats of GEHs.

Author Contributions: Conceptualization, I.C. and G.-C.C.; Methodology, I.C.; Investigation and Data duration, M.L.; Writing—original draft preparation, I.C. and M.L.; Writing—review and editing, I.C.; Supervision: G.-C.C.

Funding: The research described in this paper was financially supported by a grant from the Water Management Research Program funded by the Ministry of Land, Infrastructure, and Transport (MOLIT) of the Korean government (19AWMP-B114119-04); the National Research Foundation of Korea (NRF) grant funded by the Korea government (MSIT) (No. 2017R1A5A1014883); and the Innovated Talent Education Program for Smart City from MOLIT.

Conflicts of Interest: The authors declare no conflict of interest.

References

1. Pachauri, R.K.; Allen, M.R.; Barros, V.R.; Broome, J.; Cramer, W.; Christ, R.; Church, J.A.; Clarke, L.; Dahe, Q.; Dasgupta, P. *Climate Change 2014: Synthesis Report. Contribution of Working Groups I, II and III to the Fifth Assessment Report of the Intergovernmental Panel on Climate Change*; Intergovernmental Panel on Climate Change (IPCC): Geneva, Switzerland, 2014; p. 151.
2. Schneider, S.H. The greenhouse effect: Science and policy. *Science* **1989**, *243*, 771–781. [CrossRef] [PubMed]
3. Rodhe, H. A comparison of the contribution of various gases to the greenhouse effect. *Science* **1990**, *248*, 1217–1219. [CrossRef] [PubMed]
4. Dlugokencky, E.; Tans, P. National Oceanic & Atmospheric Administraion (NOAA) Earth System Research Laboratory (ESRL). Available online: https://www.esrl.noaa.gov/gmd/ccgg/trends/ (accessed on 7 June 2019).
5. Le Quéré, C.; Andrew, R.M.; Friedlingstein, P.; Sitch, S.; Pongratz, J.; Manning, A.C.; Korsbakken, J.I.; Peters, G.P.; Canadell, J.G.; Jackson, R.B. Global carbon budget 2017. *Earth Syst. Sci. Data* **2018**, *10*, 405–448. [CrossRef]
6. Karl, T.R.; Trenberth, K.E. Modern global climate change. *Science* **2003**, *302*, 1719–1723. [CrossRef]
7. Olivier, J.; Schure, K.; Peters, J. *Trends in Global CO_2 and Total Greenhouse Gas Emissions: 2017 Report*; PBL Netherlands Environmental Assessment Agency: Hague, The Netherlands, 2017; p. 69.
8. Goddard Institute for Space Studies (GISS) at The National Aeronautics and Space Administration (NASA). Global Land-Ocean Temperature Index. Available online: https://climate.nasa.gov/vital-signs/global-temperature/ (accessed on 7 June 2019).
9. UNFCCC. *Adoption of the Paris Agreement*; FCCC/CP/2015/L.9/Rev.1; United Nations Framework Convention on Climate Change (UNFCCC): Bonn, Germany, 2015.
10. Jackson, R.; Le Quéré, C.; Andrew, R.; Canadell, J.; Peters, G.; Roy, J.; Wu, L. Warning signs for stabilizing global CO_2 emissions. *Environ. Res. Lett.* **2017**, *12*, 110202. [CrossRef]
11. Rogelj, J.; Den Elzen, M.; Höhne, N.; Fransen, T.; Fekete, H.; Winkler, H.; Schaeffer, R.; Sha, F.; Riahi, K.; Meinshausen, M. Paris Agreement climate proposals need a boost to keep warming well below 2 °C. *Nature* **2016**, *534*, 631–639. [CrossRef] [PubMed]
12. Aalst, M.K.V. The impacts of climate change on the risk of natural disasters. *Disasters* **2006**, *30*, 5–18. [CrossRef]
13. Trenberth, K.E. Changes in precipitation with climate change. *Clim. Res.* **2011**, *47*, 123–138. [CrossRef]

14. Bouwer, L.M. Have disaster losses increased due to anthropogenic climate change? *Bull. Am. Meteorol. Soc.* **2011**, *92*, 39–46. [CrossRef]

15. Yasuhara, K.; Komine, H.; Murakami, S.; Chen, G.; Mitani, Y.; Duc, D. Effects of climate change on geo-disasters in coastal zones and their adaptation. *Geotext. Geomembr.* **2012**, *30*, 24–34. [CrossRef]

16. Mirza, M.M.Q. Climate change and extreme weather events: Can developing countries adapt? *Clim. Policy* **2003**, *3*, 233–248. [CrossRef]

17. Yasuhara, K.; Murakami, S.; Mimura, N.; Komine, H.; Recio, J. Influence of global warming on coastal infrastructural instability. *Sustain. Sci.* **2007**, *2*, 13–25. [CrossRef]

18. Mukherjee, S.; Nateghi, R.; Hastak, M. A Multi-hazard approach to assess severe weather-induced major power outage risks in the U.S. *Reliab. Eng. Syst. Saf.* **2018**, *175*, 283–305. [CrossRef]

19. National Oceanic and Atmospheric Administration (NOAA). Billion-Dollar Weather and Climate Disasters. Available online: https://www.ncdc.noaa.gov/billions/time-series (accessed on 7 June 2019).

20. Ministry of the Interior and Safety. *Annual Disaster Report 2016*; Ministry of the Interior and Safety: Seoul, Korea, 2017.

21. National Emergency Management Agency. *Annual Disaster Report 2009*; National Emergency Management Agency: Seoul, Korea, 2010.

22. National Research Council. *Geological and Geotechnical Engineering in the New Millennium: Opportunities for Research and Technological Innovation*; National Academies Press: Washington, DC, USA, 2006.

23. Chang, I.; Im, J.; Cho, G.C. Introduction of microbial biopolymers in soil treatment for future environmentally-friendly and sustainable geotechnical engineering. *Sustainability* **2016**, *8*, 251. [CrossRef]

24. Huber, M.; Knutti, R. Anthropogenic and natural warming inferred from changes in Earth's energy balance. *Nat. Geosci.* **2012**, *5*, 31–36. [CrossRef]

25. Trenberth, K.E. Atmospheric moisture residence times and cycling: Implications for rainfall rates and climate change. *Clim. Chang.* **1998**, *39*, 667–694. [CrossRef]

26. Trenberth, K.E.; Dai, A.; Rasmussen, R.M.; Parsons, D.B. The changing character of precipitation. *Bull. Am. Meteorol. Soc.* **2003**, *84*, 1205–1217. [CrossRef]

27. Liu, S.C.; Fu, C.; Shiu, C.J.; Chen, J.P.; Wu, F. Temperature dependence of global precipitation extremes. *Geophys. Res. Lett.* **2009**, *36*, L11702. [CrossRef]

28. Ren, D. *Storm-Triggered Landslides in Warmer Climates*; Springer: Basel, Switzerland, 2015.

29. Mimura, N. Sea-level rise caused by climate change and its implications for society. *Proc. Jpn. Acad. Ser. B Phys. Biol. Sci.* **2013**, *89*, 281–301. [CrossRef]

30. Church, J.A.; White, N.J.; Konikow, L.F.; Domingues, C.M.; Cogley, J.G.; Rignot, E.; Gregory, J.M.; van den Broeke, M.R.; Monaghan, A.J.; Velicogna, I. Revisiting the Earth's sea-level and energy budgets from 1961 to 2008. *Geophys. Res. Lett.* **2011**, *38*, L18601. [CrossRef]

31. Solomon, S.; Qin, D.; Manning, M.; Chen, Z.; Marquis, M.; Averyt, K.B.; Tignor, M.; Miller, H.L. *Climate Change 2013: The Physical Science Basis. Contribution of Working Group I to the Fourth Assessment Report of the Intergovernmental Panel on Climate Change*; Intergovernmental Panel on Climate Change (IPCC): Cambridge, UK; New York, NY, USA, 2007; p. 996.

32. Iverson, R.M. Landslide triggering by rain infiltration. *Water Resour. Res.* **2000**, *36*, 1897–1910. [CrossRef]

33. Lyle, R.; Lipovsky, P.; Brideau, M.-A.; Hutchinson, D. Landslides on ice-rich slopes—A geohazard in a changing climate. In Proceedings of the Geotechnical Engineering for Infrastructure and Development: XVI European Conference on Soil Mechanics and Geotechnical Engineering (ECSMGE), Edinburgh, UK, 13–17 September 2015.

34. Emanuel, K. A statistical analysis of tropical cyclone intensity. *Mon. Weather Rev.* **2000**, *128*, 1139–1152. [CrossRef]

35. Crozier, M.J.; Glade, T. Landslide hazard and risk: Issues, concepts and approach. In *Landslide Hazard and Risk*; John Wiley & Sons Ltd.: Chichester, UK, 2005; pp. 1–40.

36. Keefer, D.K.; Wilson, R.C.; Mark, R.K.; Brabb, E.E.; Brown, W.M.; Ellen, S.D.; Harp, E.L.; Wieczorek, G.F.; Alger, C.S.; Zatkin, R.S. Real-time landslide warning during heavy rainfall. *Science* **1987**, *238*, 921–925. [CrossRef] [PubMed]

37. Wieczorek, G.F.; Glade, T.; Jakob, M.; Hungr, O. Climatic factors influencing occurrence of debris flows. In *Debris-Flow Hazards and Related Phenomena*; Springer: Berlin/Heidelberg, Germany, 2007; pp. 325–362.

38. Fuchu, D.; Lee, C.; Sijing, W. Analysis of rainstorm-induced slide-debris flows on natural terrain of Lantau Island, Hong Kong. *Eng. Geol.* **1999**, *51*, 279–290. [CrossRef]

39. Arenson, L.U.; Jakob, M. Periglacial Geohazard Risks and Ground Temperature Increases. In *Engineering Geology for Society and Territory—Volume 1*; Springer International Publishing: Cham, Switzerland, 2015; pp. 233–237.

40. Highland, L.; Bobrowsky, P.T. *The Landslide Handbook: A Guide to Understanding Landslides*; U.S. Geological Survey: Reston, VA, USA, 2008.

41. Williams, P.J. Permafrost and climate change: Geotechnical implications. *Philos. Trans. R. Soc. Lond. A* **1995**, *352*, 347–358.

42. Haeberli, W.; Beniston, M. Climate change and its impacts on glaciers and permafrost in the Alps. *Ambio* **1998**, *27*, 258–265.

43. Lamoreaux, P.E.; Newton, J. Catastrophic subsidence: An environmental hazard, Shelby County, Alabama. *Environ. Geol. Water Sci.* **1986**, *8*, 25–40. [CrossRef]

44. Waltham, T.; Bell, F.G.; Culshaw, M.G. *Sinkholes and Subsidence: Karst and Cavernous Rocks in Engineering and Construction*; Springer Science & Business Media: Hidelberg, Germany, 2007.

45. Huang, W.-C.; Weng, M.-C.; Chen, R.-K. Levee failure mechanisms during the extreme rainfall event: A case study in Southern Taiwan. *Nat. Hazards* **2014**, *70*, 1287–1307. [CrossRef]

46. El Shamy, U.; Aydin, F. Multiscale modeling of flood-induced piping in river levees. *J. Geotech. Geoenviron. Eng.* **2008**, *134*, 1385–1398. [CrossRef]

47. Ojha, C.; Singh, V.; Adrian, D. Determination of critical head in soil piping. *J. Hydraul. Eng.* **2003**, *129*, 511–518. [CrossRef]

48. Williams, M. *Climate Change in Deserts*; Cambridge University Press: New York, NY, USA, 2014.

49. Zhu, L.; Gong, H.; Li, X.; Wang, R.; Chen, B.; Dai, Z.; Teatini, P. Land subsidence due to groundwater withdrawal in the northern Beijing plain, China. *Eng. Geol.* **2015**, *193*, 243–255. [CrossRef]

50. Faunt, C.C.; Sneed, M.; Traum, J.; Brandt, J.T. Water availability and land subsidence in the Central Valley, California, USA. *Hydrogeol. J.* **2016**, *24*, 675–684. [CrossRef]

51. Briaud, J.-L.; Chen, H.-C.; Govindasamy, A.; Storesund, R. Levee erosion by overtopping in New Orleans during the Katrina Hurricane. *J. Geotech. Geoenviron. Eng.* **2008**, *134*, 618–632. [CrossRef]

52. Alongi, D.M. Mangrove forests: Resilience, protection from tsunamis, and responses to global climate change. *Estuar. Coast. Shelf Sci.* **2008**, *76*, 1–13. [CrossRef]

53. Vardon, P.J. Climatic influence on geotechnical infrastructure: A review. *Environ. Geotech.* **2015**, *2*, 166–174. [CrossRef]

54. Floods and landslides leave dozens dead and 50 missing in Japan. *The Guardian.* 7 July 2018. Available online: https://www.theguardian.com/world/2018/jul/07/heavy-rain-floods-and-landslide-leave-more-than-a-dozen-dead-and-50-missing-in-japan (accessed on 7 June 2019).

55. West Japan flood victims still living in despair a month after disaster: Survey. *The Japan Times.* 6 August 2018. Available online: https://www.japantimes.co.jp/news/2018/08/06/national/west-japan-flood-victims-still-living-despair-month-disaster-survey/#.XHOYnzMzaUl (accessed on 7 June 2019).

56. Emanuel, K. Increasing destructiveness of tropical cyclones over the past 30 years. *Nature* **2005**, *436*, 686–688. [CrossRef] [PubMed]

57. Paul, S.; Ghebreyesus, D.; Sharif, H.O. Brief Communication: Analysis of the Fatalities and Socio-Economic Impacts Caused by Hurricane Florence. *Geosciences* **2019**, *9*, 58. [CrossRef]

58. The Interactive Relationship between Coastal Erosion and Flood Risk. SAGE Publications. Available online: https://journals.sagepub.com/doi/abs/10.1177/0309133318794498 (accessed on 7 June 2019).

59. Ridley, A.; McGinnity, B.; Vaughan, P. Role of pore water pressures in embankment stability. *Proc. Inst. Civ. Eng. Geotech. Eng.* **2004**, *157*, 193–198. [CrossRef]

60. Sanderson, D.; Sharma, A. *World Disasters Report 2016*; International Federation of Red Cross and Red Cresent Societies (IFRC): Geneva, Switzerland, 2016; p. 276.

61. Petley, D. Global patterns of loss of life from landslides. *Geology* **2012**, *40*, 927–930. [CrossRef]

62. Barredo, J.I. Normalised flood losses in Europe: 1970–2006. *Nat. Hazards Earth Syst. Sci.* **2009**, *9*, 97–104. [CrossRef]

63. Alcantara-Ayala, I. Geomorphology, natural hazards, vulnerability and prevention of natural disasters in developing countries. *Geomorphology* **2002**, *47*, 107–124. [CrossRef]

64. Hirabayashi, Y.; Kanae, S.; Emori, S.; Oki, T.; Kimoto, M. Global projections of changing risks of floods and droughts in a changing climate. *Hydrol. Sci. J.* **2008**, *53*, 754–772. [CrossRef]
65. Huggel, C.; Stone, D.; Auffhammer, M.; Hansen, G. Loss and damage attribution. *Nat. Clim. Chang.* **2013**, *3*, 694–696. [CrossRef]
66. Thomas, V.; López, R. Global increase in climate-related disasters. *Asian Dev. Bank Econ. Work. Pap. Ser.* 2015. Available online: http://hdl.handle.net/11540/5274 (accessed on 7 June 2019).
67. Centre for Research on the Epidemiology of Disasters (CRED). Emergency Events Database (EM-DAT). Available online: https://www.emdat.be/emdat_db/ (accessed on 7 June 2019).
68. Huang, R.; Chen, J.; Huang, G. Characteristics and variations of the East Asian monsoon system and its impacts on climate disasters in China. *Adv. Atmos. Sci.* **2007**, *24*, 993–1023. [CrossRef]
69. Haque, C.E. Perspectives of natural disasters in East and South Asia, and the Pacific Island States: Socio-economic correlates and needs assessment. *Nat. Hazards* **2003**, *29*, 465–483. [CrossRef]
70. Toya, H.; Skidmore, M. Economic development and the impacts of natural disasters. *Econ. Lett.* **2007**, *94*, 20–25. [CrossRef]
71. Raschky, P.A. Institutions and the losses from natural disasters. *Nat. Hazards Earth Syst. Sci.* **2008**, *8*, 627–634. [CrossRef]
72. Padli, J.; Habibullah, M.S.; Baharom, A.H. The impact of human development on natural disaster fatalities and damage: Panel data evidence. *Econ. Res. Ekonomska Istraživanja* **2018**, *31*, 1557–1573. [CrossRef]
73. Shen, G.; Hwang, S.N. Spatial–Temporal snapshots of global natural disaster impacts Revealed from EM-DAT for 1900–2015. *Geomat. Nat. Hazards Risk* **2019**, *10*, 912–934. [CrossRef]
74. Fragaszy, R.J.; Santamarina, J.C.; Amekudzi, A.; Assimaki, D.; Bachus, R.; Burns, S.E.; Cha, M.; Cho, G.C.; Cortes, D.D.; Dai, S.; et al. Sustainable development and energy geotechnology—Potential roles for geotechnical engineering. *KSCE J. Civ. Eng.* **2011**, *15*, 611–621. [CrossRef]
75. Espinoza, D.N.; Kim, S.H.; Santamarina, J.C. CO_2 geological storage—Geotechnical implications. *KSCE J. Civ. Eng.* **2011**, *15*, 707–719. [CrossRef]
76. Abramson, L.W. *Slope Stability and Stabilization Methods*; John Wiley & Sons: Hoboken, NJ, USA, 2002.
77. Shukla, S.; Sivakugan, N.; Das, B. A state-of-the-art review of geosynthetic-reinforced slopes. *Int. J. Geotech. Eng.* **2011**, *5*, 17–32. [CrossRef]
78. Kourkoulis, R.; Gelagoti, F.; Anastasopoulos, I.; Gazetas, G. Hybrid method for analysis and design of slope stabilizing piles. *J. Geotech. Geoenviron. Eng.* **2011**, *138*, 1–14. [CrossRef]
79. Prashant, A.; Mukherjee, M. *Soil Nailing for Stabilization of Steep Slopes Near Railway Tracks*; Department of Civil Engineering, Indian Institute of Technology Kanpur: Kanpur, India, 2010.
80. Lamont-Black, J.; Jones, C.J.F.P.; Alder, D. Electrokinetic strengthening of slopes—Case history. *Geotext. Geomembr.* **2016**, *44*, 319–331. [CrossRef]
81. Sills, G.; Vroman, N.; Wahl, R.; Schwanz, N. Overview of New Orleans levee failures: Lessons learned and their impact on national levee design and assessment. *J. Geotech. Geoenviron. Eng.* **2008**, *134*, 556–565. [CrossRef]
82. Alfaro, M.; Balasubramaniam, A.; Bergado, D.; Chai, J. *Improvement Techniques of Soft Ground in Subsiding and Lowland Environment*; CRC Press: Boca Raton, FL, USA, 1994.
83. Jan, V.H. The use of polyurethane in coastal engineering models. In Proceedings of the 5th Conference of the Application of Physical Modelling to Port and Coastal Protection (Coastlab14), Varna, Bulgaria, 29 September–2 October 2014; Volume 2, pp. 178–185.
84. Andrew, R.M. Global CO_2 emissions from cement production, 1928–2017. *Earth Syst. Sci. Data* **2018**, *10*, 2213–2239. [CrossRef]
85. United States Geological Survey (USGS). Cement Statistics and Information. Available online: https://www.usgs.gov/centers/nmic/cement-statistics-and-information (accessed on 7 June 2019).
86. Tingle, J.; Newman, J.; Larson, S.; Weiss, C.; Rushing, J. Stabilization mechanisms of nontraditional additives. *Transp. Res. Rec. J. Transp. Res. Board* **2007**, *1989*, 59–67. [CrossRef]
87. Santoni, R.; Tingle, J.; Webster, S. Stabilization of Silty Sand with Nontraditional Additives. *Transp. Res. Rec. J. Transp. Res. Board* **2002**, *1787*, 61–70. [CrossRef]
88. Green, V.S.; Stott, D. Polyacrylamide: A review of the use, effectiveness, and cost of a soil erosion control amendment. In Proceedings of the 10th International Soil Conservation Organization Meeting, Purdue, IN, USA, 24–29 May 1999; pp. 384–389.

89. Marto, A.; Yunus, M.; Zurairahetty, N.; Pakir, F.; Latifi, N.; Nor, M.; Hakimi, A.; Tan, C.S. Stabilization of marine clay by biomass silica (non-traditional) stabilizers. *Appl. Mech. Mater.* **2015**, *695*, 93–97. [CrossRef]

90. Latifi, N.; Meehan, C.L. Strengthening of montmorillonitic and kaolinitic clays with calcium carbide residue: A sustainable additive for soil stabilization. In Proceedings of the Geotechnical Frontiers 2017, Orlando, FL, USA, 12–15 May 2017; pp. 154–163.

91. Onyejekwe, S.; Ghataora, G.S. Soil stabilization using proprietary liquid chemical stabilizers: Sulphonated oil and a polymer. *Bull. Eng. Geol. Environ.* **2015**, *74*, 651–665. [CrossRef]

92. Karol, R.H. *Chemical Grouting and Soil Stabilization, Revised and Expanded*; CRC Press: Boca Raton, FL, USA, 2003; Volume 12.

93. Ding, Y.; Dai, J.-G.; Shi, C.-J. Mechanical properties of alkali-activated concrete: A state-of-the-art review. *Constr. Build. Mater.* **2016**, *127*, 68–79. [CrossRef]

94. Saravanan, G.; Jeyasehar, C.; Kandasamy, S. Flyash Based Geopolymer Concrete-A State of the Art Review. *J. Eng. Sci. Technol. Rev.* **2013**, *6*, 25–32. [CrossRef]

95. Toniolo, N.; Boccaccini, A.R. Fly ash-based geopolymers containing added silicate waste. A review. *Ceram. Int.* **2017**, *43*, 14545–14551. [CrossRef]

96. Singh, B.; Ishwarya, G.; Gupta, M.; Bhattacharyya, S.K. Geopolymer concrete: A review of some recent developments. *Constr. Build. Mater.* **2015**, *85*, 78–90. [CrossRef]

97. Phetchuay, C.; Horpibulsuk, S.; Arulrajah, A.; Suksiripattanapong, C.; Udomchai, A. Strength development in soft marine clay stabilized by fly ash and calcium carbide residue based geopolymer. *Appl. Clay Sci.* **2016**, *127–128*, 134–142. [CrossRef]

98. Cristelo, N.; Glendinning, S.; Teixeira Pinto, A. Deep soft soil improvement by alkaline activation. *Proc. Inst. Civ. Eng. Ground Improv.* **2011**, *164*, 73–82. [CrossRef]

99. Duxson, P.; Provis, J.L.; Lukey, G.C.; van Deventer, J.S.J. The role of inorganic polymer technology in the development of 'green concrete'. *Cem. Concr. Res.* **2007**, *37*, 1590–1597. [CrossRef]

100. Provis, J.L.; Van Deventer, J.S.J. *Geopolymers: Structures, Processing, Properties and Industrial Applications*; Elsevier: Amsterdam, The Netherlands, 2009.

101. Zhang, M.; Guo, H.; El-Korchi, T.; Zhang, G.; Tao, M. Experimental feasibility study of geopolymer as the next-generation soil stabilizer. *Constr. Build. Mater.* **2013**, *47*, 1468–1478. [CrossRef]

102. Zornberg, J.G.; Christopher, B.R. Chapter 37: Geosynthetics. In *The Handbook of Groundwater Engineering*, 2nd ed.; Delleur, J.W., Ed.; CRC Press, Taylor & Francis Group: Boca Raton, FL, USA, 2007; Volume 2.

103. Holtz, R.D. *Geosynthetics for Soil Reinforcement: The 9th Spencer J. Buchanan Lecture*; College Station Hilton: Seattle, WA, USA, 2001.

104. Müller, W.W.; Saathoff, F. Geosynthetics in geoenvironmental engineering. *Sci. Technol. Adv. Mater.* **2015**, *16*, 034605. [CrossRef]

105. Haider, N.; Karlsson, S. Loss of Chimassorb 944 from LDPE and identification of additive degradation products after exposure to water, air and compost. *Polym. Degrad. Stab.* **2001**, *74*, 103–112. [CrossRef]

106. Beißmann, S.; Stiftinger, M.; Grabmayer, K.; Wallner, G.; Nitsche, D.; Buchberger, W. Monitoring the degradation of stabilization systems in polypropylene during accelerated aging tests by liquid chromatography combined with atmospheric pressure chemical ionization mass spectrometry. *Polym. Degrad. Stab.* **2013**, *98*, 1655–1661. [CrossRef]

107. Hatami, K.; Grady, B.P.; Ulmer, M.C. Sensor-enabled geosynthetics: Use of conducting carbon networks as geosynthetic sensors. *J. Geotech. Geoenviron. Eng.* **2009**, *135*, 863–874. [CrossRef]

108. Raisinghani, D.V.; Viswanadham, B.V.S. Centrifuge model study on low permeable slope reinforced by hybrid geosynthetics. *Geotext. Geomembr.* **2011**, *29*, 567–580. [CrossRef]

109. Mujah, D.; Shahin, M.A.; Cheng, L. State-of-the-art review of biocementation by microbially induced calcite precipitation (MICP) for soil stabilization. *Geomicrobiol. J.* **2017**, *34*, 524–537. [CrossRef]

110. DeJong, J.T.; Soga, K.; Banwart, S.A.; Whalley, W.R.; Ginn, T.R.; Nelson, D.C.; Mortensen, B.M.; Martinez, B.C.; Barkouki, T. Soil engineering in vivo: Harnessing natural biogeochemical systems for sustainable, multi-functional engineering solutions. *J. R. Soc. Interfaces* **2010**, *8*, 1–15. [CrossRef] [PubMed]

111. Neupane, D.; Yasuhara, H.; Kinoshita, N.; Unno, T. Applicability of Enzymatic Calcium Carbonate Precipitation as a Soil-Strengthening Technique. *J. Geotech. Geoenviron. Eng.* **2013**, *139*, 2201–2211. [CrossRef]

112. Whiffin, V.S.; van Paassen, L.A.; Harkes, M.P. Microbial carbonate precipitation as a soil improvement technique. *Geomicrobiol. J.* **2007**, *24*, 417–423. [CrossRef]

113. Chu, J.; Ivanov, V.; He, J.; Maeimi, M.; Wu, S. Chapter 19—Use of Biogeotechnologies for Soil Improvement. In *Ground Improvement Case Histories*; Elsevier: Amsterdam, The Netherlands, 2015; pp. 571–589.

114. Fujita, Y.; Taylor, J.L.; Gresham, T.L.T.; Delwiche, M.E.; Colwell, F.S.; McLing, T.L.; Petzke, L.M.; Smith, R.W. Stimulation of microbial urea hydrolysis in groundwater to enhance calcite precipitation. *Environ. Sci. Technol.* **2008**, *42*, 3025–3032. [CrossRef]

115. Gomez, M.G.; Martinez, B.C.; DeJong, J.T.; Hunt, C.E.; deVlaming, L.A.; Major, D.W.; Dworatzek, S.M. Field-scale bio-cementation tests to improve sands. *Proc. Inst. Civ. Eng. Ground Improv.* **2015**, *168*, 206–216. [CrossRef]

116. Cheng, L.; Shahin, M.; Cord-Ruwisch, R. Bio-cementation of sandy soil using microbially induced carbonate precipitation for marine environments. *Geotechnique* **2014**, *64*, 1010–1013. [CrossRef]

117. Achal, V.; Mukherjee, A.; Kumari, D.; Zhang, Q. Biomineralization for sustainable construction–A review of processes and applications. *Earth Sci. Rev.* **2015**, *148*, 1–17. [CrossRef]

118. Maher, M.; Ho, Y. Mechanical Properties of Kaolinite/Fiber Soil Composite. *J. Geotech. Eng.* **1994**, *120*, 1381–1393. [CrossRef]

119. Chang, I.; Shin, Y.; Cho, G. Optimum thickness decision of biopolymer treated soil for slope protection on the soil slope. In Proceedings of the 14th International Conference of International Association for Computer Methods and Recent Advances in Geomechanics, Kyoto, Japan, 22–25 September 2014; pp. 1643–1648.

120. Chang, I.; Im, J.; Prasidhi, A.K.; Cho, G.-C. Effects of Xanthan gum biopolymer on soil strengthening. *Constr. Build. Mater.* **2015**, *74*, 65–72. [CrossRef]

121. Chang, I.; Prasidhi, A.K.; Im, J.; Cho, G.-C. Soil strengthening using thermo-gelation biopolymers. *Constr. Build. Mater.* **2015**, *77*, 430–438. [CrossRef]

122. Chang, I.; Jeon, M.; Cho, G.-C. Application of microbial biopolymers as an alternative construction binder for earth buildings in underdeveloped countries. *Int. J. Polym. Sci.* **2015**, *2015*, 326745. [CrossRef] [PubMed]

123. Chang, I.; Prasidhi, A.K.; Im, J.; Shin, H.-D.; Cho, G.-C. Soil treatment using microbial biopolymers for anti-desertification purposes. *Geoderma* **2015**, *253–254*, 39–47. [CrossRef]

124. Chang, I.; Im, J.; Cho, G.-C. Geotechnical engineering behaviors of gellan gum biopolymer treated sand. *Can. Geotech. J.* **2016**, *53*, 1658–1670. [CrossRef]

125. Chang, I.; Im, J.; Lee, S.-W.; Cho, G.-C. Strength durability of gellan gum biopolymer-treated Korean sand with cyclic wetting and drying. *Constr. Build. Mater.* **2017**, *143*, 210–221. [CrossRef]

126. Chang, I.; Im, J.; Chung, M.-K.; Cho, G.-C. Bovine casein as a new soil strengthening binder from diary wastes. *Constr. Build. Mater.* **2018**, *160*, 1–9. [CrossRef]

127. Survase, S.A.; Saudagar, P.S.; Bajaj, I.B.; Singhal, R.S. Scleroglucan: Fermentative production, downstream processing and applications. *Food Technol. Biotechnol.* **2007**, *45*, 107–118.

128. Chang, I.; Cho, G.-C. Strengthening of Korean residual soil with β-1,3/1,6-glucan biopolymer. *Constr. Build. Mater.* **2012**, *30*, 30–35. [CrossRef]

129. Bajaj, I.B.; Survase, S.A.; Saudagar, P.S.; Singhal, R.S. Gellan gum: Fermentative production, downstream processing and applications. *Food Technol. Biotechnol.* **2007**, *45*, 341–354.

130. Reddy, N.; Reddy, R.; Jiang, Q. Crosslinking biopolymers for biomedical applications. *Trends Biotechnol.* **2015**, *33*, 362–369. [CrossRef] [PubMed]

energies

MDPI

Article

Effect of Clay Content on the Mechanical Properties of Hydrate-Bearing Sediments during Hydrate Production via Depressurization

Dongliang Li [1,2], Zhe Wang [1,2,3], Deqing Liang [1,2,*] and Xiaoping Wu [4,5]

[1]   Key Laboratory of Gas Hydrate, Guangzhou Institute of Energy Conversion, Chinese Academy of Sciences, Guangzhou 510640, China
[2]   Guangdong Provincial Key Laboratory of New and Renewable Energy Research and Development, Guangzhou 510640, China
[3]   Nano Science and Technology Institute, University of Science and Technology of China, Suzhou 215123, China
[4]   School of Earth & Space Science, University of Science and Technology of China, Hefei 230026, China
[5]   CAS Center for Excellence in Comparative Planetology, Hefei 230026, China
*   Correspondence: liangdq@ms.giec.ac.cn; Tel.: +86-20-8705-7657; Fax: +86-20-8705-7669

Received: 11 June 2019; Accepted: 11 July 2019; Published: 12 July 2019

Abstract: The effects of sediments with different clay contents on the mechanical properties of hydrate deposits were studied using a high-pressure, low-temperature triaxial apparatus with in-situ synthesis, as well as the mechanical properties of self-developed hydrate sediments. Through multi-stage loading, triaxial compression tests were conducted by adding quartz sand with different clay contents as the sediment skeleton, and the stress–strain relationship of the shearing process and the strength of sediments with different clay contents were determined. Volumetric changes were also observed during shearing. The results show that the strength of hydrate sediments decreases with the increasing clay content of sediments; in the processes of depressurization and shearing, the hydrate samples exhibited obvious shear shrinkage, regardless of the sediment particle size.

Keywords: triaxial shear; methane hydrate; clay content; mechanical property; hydrate mining; shear shrinkage

1. Introduction

The demand for energy in various countries is increasing with the continued development of modern society. The resources of traditional fossil fuels have diminished due to years of exploitation and utilization, and the world is facing an increasingly severe energy crisis. The natural gas hydrates that have been discovered thus far offer potential energy sources with which fossil fuels may be replaced. Natural gas hydrates are generally formed in low-temperature and high-pressure environments, and are mainly stored on deep-sea slopes and in permafrost regions [1,2]. Gas hydrates, primarily composed of methane, are naturally distributed across various regions worldwide. Because their shapes are similar to those of ice, they are often called "combustible ice." According to one survey, the reserves of natural gas hydrates may be 2.1×10^{16} m^3, and the decomposition of methane hydrates at approximately one atmosphere can produce ~160 m^3 of methane gas and ~0.87 m^3 of free water [3]. It is estimated that this is roughly twice as much as all carbonaceous fossil fuel reserves in the world [4].

Methane hydrate has attracted worldwide attention due to its wide distribution, large scale development, and high energy storage density. Extensive research on hydrates has been conducted in China, the United States, Japan, Canada, and other countries. At present, the methane hydrate mining methods proposed primarily include injection [5], pressure reduction [6], chemical reagents, and CO_2 replacement [7]. During the mining process, artificially breaking the stable conditions of the hydrate will cause it to decompose. Studies have shown that hydrates are cemented between sediments, which enhances the strength of seafloor sediments. In the process of hydrate mining, the hydrate will gradually decompose, the cementation and filling of the hydrate will be reduced, and the bearing capacity of the bottom layer will be greatly reduced. This may cause a series of geological disasters, if occurring at the bottom of the sea, including large tsunami and submarine landslides [8,9]. Therefore, it is very important to study the mechanical properties of hydrate-bearing sediments.

Since methane hydrates are mainly distributed in the deep sea and in permafrost regions, the cost of obtaining in-situ hydrated sediment cores for research is high and poses technical difficulties. Therefore, laboratory-synthesized gas hydrates are currently used as samples to study the relevant properties and various parameters of natural hydrates. The mechanical properties of hydrate sediments are mainly affected by hydrate saturation, confining pressure, sediment particle size, and sediment type. Researchers worldwide have performed studies on some of these properties. Winters et al. [10,11] used stored sand to study the effects of sediment type and porosity on hydrate sediments., and Hyodo et al. [12–14] used in-situ synthesis and mixed sample preparation methods to study the sedimentation, temperature, hydrate formation states (e.g., gas saturation formation, water saturation generation), and other hydrate formations in different sediments. The influence of mechanical properties indicates that the strength of hydrate sediments increases with the increase of hydrate saturation and confining pressure, and the strength of gas-saturated hydrate sediments is higher than that of other hydrate sediments under the same conditions. Moreover, Li et al. [15,16] studied the influence of hydrates on confining pressure and noted that the strength of hydrate sediments increased with the increase of confining pressure under certain conditions. Meanwhile, Song et al. [17,18] studied the effects of different temperatures and confining pressures on hydrate sediments. It is believed that under certain conditions, an increase in confining pressure will lead to an increase in the strength of hydrate sediments, as well as increases in temperature and shear rate, leading to an increase in shear strength.

Masui et al. [19] analyzed natural gas hydrates obtained from the South China Sea Trough in 2004 and synthesized matching hydrate sediment samples in the laboratory, with the original particle distribution, and distributed the two sets of samples for triaxial compression experiments. They concluded that the strengths of the natural and laboratory-synthesized hydrate sediments were consistent, and that the volume deformation characteristics were also the same. In order to study the effects of sand particle size on the mechanical properties of natural gas hydrate sediments, Miyazaki et al. [20] used Toyoura sand with three particle sizes as hydrate sediments, with median particle sizes (d50) of 0.230 mm, 0.205 mm, and 0.130 mm. The material skeleton was subjected to a drainage triaxial compression experiment under a constant temperature of 287 K. Their results showed that the strength of methane hydrate sediments increases with the saturation of hydrates and the effective confining pressure. Moreover, the strength of the hydrate sediments had little impact on the change in sand particle sizes, and the stiffness of the hydrate sediments was affected by the sand skeleton.

Kajiyama et al. [21] used circular glass beads and natural sand as the skeleton of methane hydrate deposits, and then performed a series of triaxial compression experiments on the two hydrate deposits to study their mechanical properties. The effects of particle characteristics on the mechanical properties of the methane-bearing hydrate sand were explained from the perspective of particle size. The stiffness of the glass bead skeleton and the reservoir sand was relatively uniform, but the maximum breaking strength was achieved in a short period of time. The apparent post-peak-strain softening behavior of the glass beads containing methane hydrate was observed experimentally. The main reason for the increase in the shear strength of the hydrate deposit with the glass bead skeleton was cohesion, but the shear strength of the natural sand containing methane hydrate was jointly controlled by the cohesive force and the internal angle of friction. This caused the strength of the methane hydrate deposit in the natural sand skeleton to increase with the increase of the effective confining pressure, while the methane hydrate deposit in the glass bead skeleton had greater cohesion under relatively low pressures. In strong sediments and at higher effective confining pressures, cohesion will be greatly reduced due to the detachment of hydrates.

Hyodo et al. [14,22] used a self-made hydrate triaxial compression instrument to synthesize methane hydrate sediments using three grits with different densities as the methane sediment skeleton and by adding different contents of fine particles to the sand. They concluded that the addition of fine particles had a significant effect on the porosity of the sedimentary skeleton, and the fine particles filled in between the hydrate sediment skeletons, making the sample more compact and the initial porosity of the sample lower. Due to this lower porosity, the methane hydrate formed more severely hindered the movement of the particles, thereby increasing the strength of the entire methane hydrate sample. The porosity of the hydrate sediment samples was inversely proportional to the percentage of fine particles, and the strength of the hydrate sediments also increased. Additionally, the effects of fine particles on pure sand sediments and methane hydrate sediments differed. The presence of fine particles increased the strength of methane hydrate sediments and methane hydrates during triaxial shearing; meanwhile, dilatation occurred in the sediments, and pure sand sediments underwent shearing.

Many researchers have studied the mechanical properties of hydrate sediments. Most of this research has been focused on understanding the effects of sediments under single-variable conditions. However, fine-grained sediments host more than 90% of the accumulated global gas hydrates [23–25]; these accumulations of hydrate-bearing clayey sediments include those in the Gulf of Mexico, Krishna–Godavari Basin, Blake Ridge, Cascadia Margin, Ulleung Basin, Hydrate Ridge [26], and South China Sea [27]. Therefore, it is necessary to study the influence of clay content on the mechanical properties of sediments under multifactorial conditions. In this study, quartz sands with different proportions of clay were used as sediment skeletons, and the effects of clay content on the mechanical properties of hydrate sediments were studied using multi-stage loading triaxial compression tests combined with hydrate decomposition [28].

2. Materials and Methods

2.1. Experimental Apparatus

Figure 1 shows a schematic of the low-temperature, high-pressure triaxial apparatus used for the in-situ synthesis and mechanical property testing of self-developed hydrate sediments. The axial loading pressure ranged from 0–250 kN, the adjustable loading rate was 0.001–6 mm/min, and the three-axis pressure chamber (i.e., hydrate synthesis reactor) could set the confining pressure and pore pressure, ranging from 0–30 MPa. The sample size was Ø = 50 mm × 100 mm, the system operating temperatures were −30–50 °C (±0.5 °C), and the temperature range of the triaxial pressure chamber was approximately −30–50 °C (±0.5 °C). The device was mainly composed of a reaction kettle, a film-forming sediment preparation system, a stress loading system, a temperature control system, a confining pressure loading system, and a vacuum system. The temperature and pressure data during the experiment were monitored via computer in real time. The data acquisition rate of the computer reached up to 6 times per second, which could accurately monitor the temperature and pressure changes.

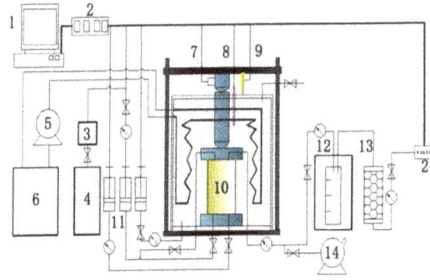

Figure 1. Schematic of the triaxial shear test apparatus: (**1**) computer; (**2**) data acquisition system; (**3**) buffer tank; (**4**) methane gas bottle; (**5**) water pump; (**6**) thermal control pump; (**7**) stress transducer; (**8**) temperature sensor; (**9**) displacement transducer; (**10**) specimen; (**11**) syringe pump; (**12**) gas–liquid separator; (**13**) desiccant; (**14**) vacuum pump [29].

2.2. Samples

The sand used in this experiment was a natural sand from the South China Sea. After screening with a 40–60 mesh sieve, deionized water was used to remove impurities, such as mud, ash, and salt. After preparation, the median grain size of the sand (d50) was found to be 377 µm, and the porosity was 33%; the grain size distribution of the sand is shown in Figure 2. The deionized water used in the experiment was made in the laboratory. The experimental clay was kaolin, for which the molecular formula is $Al_2Si_2O_9H_4$, and the relative molecular mass is 258.16.

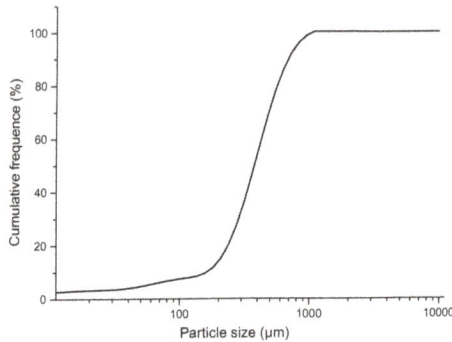

Figure 2. Grain size distribution curve.

2.3. Experimental Methods

Our experiments involved a multi-stage triaxial compression method [28], which is a repetitive loading triaxial compression shear test in soil mechanics. The same samples were used in the experiments to perform multi-stage shearing by changing the relevant parameters of the experiment. This method can effectively improve the experimental efficiency, as well as change the mechanical properties of hydrate sediment samples, in the case of changes in multiple experimental parameters.

2.3.1. Synthesis and Preparation of Samples

The sand used in the experiments was sieved and washed with deionized water, before being placed in an oven and allowed to stand at 104 °C for 10 h. After drying, a standardized amount of sand was placed into a plastic container, then the weighed clay was added in batches, and stirred simultaneously. After the mixture was almost uniform, the deionized water was added and stirred until the agglomerated particles were invisible. The sand–clay mixture was then left to stand under closed conditions for 24 h, so that the deionized water was evenly distributed among the sand grains. A rubber mold with a thickness of 0.8 mm was fixed to the base of the test bench, and then a water-permeable plate, a stainless steel metal mesh, and a fast filter paper were sequentially placed at the bottom. The metal mold was fixed outside of the rubber mold with a rubber band. The prepared sand sample was layered into the mold and compacted by a compactor. The filter paper, metal mesh, and water-permeable plate were then placed in the upper layer of the sand sample in-turn. Next, a two-way air intake cover was installed, and the sample was evacuated by a pump to detect the airtightness of the device. Afterward, the metal mold was removed and finally, the pressure chamber was closed with screws.

Water was injected into the confining pressure chamber, which slowly increased the pressure in the chamber via the booster pump, and simultaneously introduced methane gas into the sample chamber by means of the upper and lower two-way air intakes. This ensured the entire supercharging process. The pressure was always higher in the pressure chamber than in the sample chamber (i.e., the confining pressure was ~1.5–2.0 MPa higher than the pore pressure). Finally, the confining pressure was increased to 10 MPa, and the pore pressure was stopped at 9.0 MPa. The methane gas source was turned off, the sample chamber was connected to the buffer tank, the temperature of the system was maintained at 20 °C by a circulating water bath, and the sample was allowed to stand for 24 h to allow for sufficient inhalation. Thereafter, the temperature was adjusted to 5.0 °C for 48 h, before finally being adjusted to 2.0 °C for 48 h. The pore pressure of the sample was observed to judge whether or not the hydrate was completely formed.

After the hydrate had completely formed, the valve connected to the buffer tank in the sample chamber was closed. Low-temperature deionized water at ~2 °C was injected into the sample from the bottom of the sample chamber by a syringe pump, thereby draining the excess methane gas in the experiment, and changing the sample from a gas-saturated state to a water-saturated state. The remaining amount of free gas in the sample was judged by observing the gas–liquid separator at the bottom as the liquid content when the water discharge rate in the gas–liquid separator was essentially the same as the injection rate of the injection pump, and the volume of deionized water was used. When the pore volume of the sample had approximately doubled, the free gas was completely driven off by deionized water. In the process of water flooding, it was necessary to ensure that the pore pressure was lower than the experimentally required pore pressure of ~2.0 MPa, and that the pressure of the deionized water in the syringe pump was higher than the experimentally required pore pressure of ~1.0 MPa, but lower than the confining pressure. There was a pressure difference between the deionized water and the sample, such that the deionized water could be injected directly. After the water flooding process was completed, the pressure of the sample was increased using the syringe pump to the pressure required for the experiment, and the temperature was raised to 6.0 °C for the shear test.

2.3.2. Shearing and Decompression of Samples

The stress loading system was turned on and the shear rate was set to 0.20% per minute. The triaxial shearing process was divided into three stages. In the first stage, the axial strain rate was sheared from 0% to 2.2%. This stage represented undrained shear. When the axial strain rate reached ~2.2%, the axial load was unloaded and the shearing was stopped. Finally, the pore pressure was ~4.5 MPa, and liquid could be observed in the gas–liquid separator. Since the phase equilibrium pressure of methane hydrate at 6 °C is 4.73 MPa, the pore pressure of the sample was kept at 8.0 MPa during the shearing process in this stage, and the phase equilibrium condition was not destroyed. Therefore, the hydrate did not substantially decompose, and thus the amount of methane gas collected in the gas–liquid separator was extremely small. In this first stage, the pore pressure was finally reduced to ~4.5 MPa. Under the condition that the pore pressure was maintained at ~4.5 MPa and the temperature was 6.0 °C, the hydrate began to decompose slowly, and the methane gas (G_1) that decomposed after depressurization was collected and recorded.

In the second stage of shearing, the loading rate of the axial load remained consistent with the first stage, and the sample was retested under the new experimental parameters. The temperature during the shearing process was still 6.0 °C, and the pore pressure was ~4.5 MPa. In this stage, the axial strain rate was sheared from 2.2% to ~4.7%. The pore pressure was lower than with the phase equilibrium pressure of methane hydrate sediments at 6 °C, and the methane hydrate gradually began to decompose. However, since the pore pressure was slightly lower than the phase equilibrium pressure, the decomposition rate was slower. This stage still represented undrained shear. When the axial strain rate reached ~4.7%, the shearing was stopped, and the stress loading system was closed to unload the axial load. When the axial load was completely unloaded, the pressure relief valve that controlled the pore pressure was opened, and the pore pressure of the sample was gradually decreased to ~3.0 MPa. The pressure was much lower than the phase equilibrium pressure at 6.0 °C. Under the temperature and pressure conditions in this stage, the hydrate began to decompose in large amounts, and the collection rate of methane gas was much higher than in the first stage. After being decomposed into a gas, the pressure relief valve was closed, and the volume (G_2) of the methane gas generated by decomposition after pressure reduction was collected and recorded. The pressure relief valve was then closed, the axial load loading system was restored, and the third stage of shear testing was ready.

The third stage involved shearing of the axial strain rate from 4.7% to ~15%. The pore pressure at this stage was maintained at ~3.0 MPa, the system temperature was 6.0 °C, and the shear mode remained undrained. Since the pore pressure of the sample at this stage was much lower than the phase equilibrium pressure of the methane hydrate sediments at 6.0 °C, the methane hydrate in the sample was largely decomposed and the decomposition rate was faster than in previous stages. When the shear strain reached the axial strain rate of 15%, the shearing was stopped, the axial load was unloaded, the pressure relief valve was slowly opened, and the confining pressure and pore pressure of the system were gradually reduced to normal. The temperature also rose to 20 °C. After the methane hydrate in the hydrate sediments was completely decomposed and allowed to stand for one hour, the methane gas volume (G_0) generated by decomposition in all of the steps was collected and recorded; with this, the entire cutting process was complete. The difference between G_0 and G_2 is the sum of the remaining free gas and the gas generated after the decomposition of the remaining methane hydrate after the last stage. The specific experimental parameters of these tests are shown in Table 1.

Table 1. Sample parameters for shear experiments (Nos. 1–5).

Experiment No.	Clay Content (wt%)	Confining Pressure (MPa)	Pore Pressure (MPa)			Temperature (°C)	Saturation (%)		
			Stage 1	Stage 2	Stage 3		Stage 1	Stage 2	Stage 3
1	0	10	8	4.5	3	6	54.02	44.48	34.01
2	5	10	8	4.5	3	6	51.86	41.72	31.81
3	10	10	8	4.5	3	6	52.84	41.10	29.46
4	15	10	8	4.5	3	6	51.00	39.16	29.76
5	20	10	8	4.5	3	6	52.09	41.95	31.68

2.3.3. Calculation of Hydrate Saturation

Since the structure of natural gas hydrate is a non-metering cage structure, there is no way to express methane hydrate using a single, strict chemical formula, and thus the volume of methane hydrate containing a certain amount of methane gas will be not uniform. In this study, the saturation of methane hydrate (i.e., the methane hydrate in natural gas hydrate sediments) was calculated after the methods of Ghiassian et al. [30]. The theoretical basis of this method involves the assumption that a complete decomposition of methane hydrate per unit volume (1 m^3) releases ~160 units (m^3) of methane gas and an ~0.87 unit volume (m^3) of water. It is within this theoretical framework that the measurement of hydrate was considered, wherein the unit could be an ideal hydrate crystal. The specific formulae for this calculation can be seen in previous works [30].

3. Results and Discussion

3.1. Stress–Strain Relationship of Sediments with Different Clay Contents

The first stage shown in the curves of Figure 3 is the portion of the experiments with axial strain rates of 0–2.2%. At this stage, the stress–strain relationship was obtained when the sample pore pressure was 8 MPa and the confining pressure was 10 MPa (i.e., the effective confining pressure was 2 MPa). Since the pressure of the hydrate sediment sample at this stage was higher than the phase equilibrium pressure at the same temperature, the hydrate decomposition was minimal, and therefore the hydrate saturation of the sample was the highest. The stress–strain relationship at this stage reflects the initial stage of hydrate production.

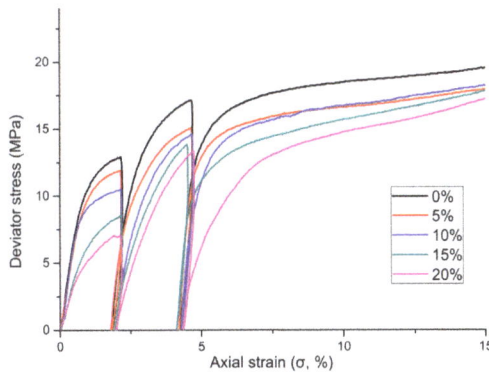

Figure 3. Stress–strain curves by clay content (%). The first stage represents axial strain rates of 0–2.2%, the middle stage accounts for rates of 2.2–4.7%, and the final arc represents strain rates of 4.7–15%. Different colors correspond to different clay contents.

It can be observed from Figure 3 that in the first stage of shearing, the stress–strain relationship between pure sand and sediments with different clay contents gradually changed from that of elastic to plastic strain. The samples initially exhibited elastic strain, indicating that they were not destroyed at the onset of shearing; when the applied stress was removed, the specimen could be restored to the original stress state. The reason for this may be that because the hydrate had not been decomposed, there were interactions between the sediment particles and the hydrate. Furthermore, because the effective confining pressure was low, the sediment particles were complete, and the particle skeleton did not change. Afterward, the elastic strain gradually changed to plastic strain, which indicates that the sample was broken, such that even if the axial load applied to the sample was removed, the sample could not be recovered. This may be due to some of the hydrate sediment particles moving under the action of the axial load and confining pressure, resulting in the cementation of the hydrate to the skeleton being destroyed and the hydrate sediment changing. At this stage, the hydrate was minimally decomposed, but due to the axial load, some of the hydrate was split, and these destroyed hydrates filled in between the hydrate sediment skeletons, making the entire sample more compact. Thus, the stress-strain relationship changed from one of elastic to plastic strain owing to changes in the skeleton and the failure of the destroyed hydrate particles. The first stage of shearing may be considered to have extended until an axial strain rate of ~2.7%, which was the maximum shear strength of the first stage. It can also be observed that the stress–strain relationships of sediments with different clay contents in the first stage differed, and the degree of plasticity gradually increased as the proportion of clay in the sediments increased. Additionally, as the clay content in the sediments increased, the strength of the hydrate sediments decreased.

The second stage shown in the curves of Figure 3 is for the experiments with axial strain rates of 2.2–4.7%. In this stage, the pore pressure was reduced from 8.0 MPa to 4.5 MPa, and the axial load was unloaded. After the hydrate in the sample was decomposed, the axial load system was restarted, and the effective confining pressure was kept at 10 MPa. Meanwhile, the stress–strain relationship was obtained by shearing under a confining pressure of 5.5 MPa. During the shearing process, the pressure of the hydrate sediment sample was slightly lower than the phase equilibrium pressure of the methane hydrate at 6.0 °C, and the hydrate also underwent partial decomposition. Since the hydrate had decomposed and exhaled part of the methane gas before shearing at this stage, the hydrate saturation of the sediment sample was lower than in the first stage of the experiments. In the second stage, during the exploitation of hydrates, the effective confining pressure of the hydrate sediments was greatly increased by depressurization, as the depressurized hydrate began to decompose.

The curve of the second stage represents the transformation from elastic to plastic strain (Figure 3). It can be observed from the stress–strain relationships of the second stage that the curve representing the same clay content had a substantially higher magnitude than during the first stage, which indicates that the strength of the second stage hydrate sediment was greater than that of the first stage. Since part of the water and methane gas in the sample were removed during the depressurization process before shearing, the hydrate sediment sample changed from a completely water-saturated state to an incompletely water-saturated state. The saturation was also lower than in the first stage. Under the same experimental conditions, the strength of water-saturated sediments was lower than that of the gas-saturated sediments, and as the abundance of hydrate sediments decreased, the methane hydrate sediments gradually decreased. The reason for this behavior may be that methane hydrate was formed between the sediment skeleton, and cemented the quartz sand skeleton, resulting in the higher strength of the methane hydrate sediment.

In the second stage of the experiments, the pore pressure of the hydrate sediments was greatly reduced, resulting in the effective confining pressure of the sample being much higher than in the first stage. Within a certain range of effective confining pressures, the strength of natural gas hydrate sediments will gradually increase with the increase of effective confining pressure. Based on the influencing factors described, it can be determined that for the methane hydrate sediments with different clay contents, the effects of the second-stage effective confining pressure and the change of hydrate sediments from water-saturated to partially water-saturated were less than that of the decreasing of hydrate saturation. More importantly, when comparing the curves of the first and second stages for the same clay content, it was found that the degree of elastic strain in the second stage was more obvious than that of plastic strain, and as the particle size of the sediments increased, the strength of the hydrate sediments also increased.

The third stage in the curves shown in Figure 3 corresponds to the part of the experiments with axial strain rates ranging from 4.7–15%, which reduced the pore pressure of the sample from 4.5 MPa to 3.0 MPa, and the effective confining pressure of this stage was 7.0 MPa. The resulting stress–strain relationship was obtained by shearing. The stress–strain relationship at this stage mainly changed from elastic to plastic strain, and finally to strain yielding. During the shearing process, the pressure of the hydrate sediment sample was much lower than the phase equilibrium pressure of the methane hydrate at 6.0 °C, at which time the hydrate in the sample was decomposed in large quantities and at a relatively fast rate. The hydrate saturation in the third stage was lower than in the first two stages, and a large volume of free water produced by the decomposition of the hydrate was also discharged. This stage simulated the middle and late stages of natural hydrate production. The hydrate had already been decomposed by a large amount, the gas hydrate sediment reservoir was greatly increased by the surrounding pressure, and the saturation of the hydrate in the reservoir was substantially reduced, resulting in hydrate sedimentation. The porosity of the sample also increased. As with the second stage, the hydrate saturation of the hydrate sediment sample was reduced due to the decomposition of the hydrate, and the decrease in the pore pressure caused the effective confining pressure of the sample to increase by 1.5 MPa relative to the second stage. However, comparing the curves of the second and third stages, it can be seen that the curve in which the axial strain rate changed to 2.5% in the third stage is essentially the same as the curve in the second stage, which indicates that the effective confining pressure increased in this stage, yet the effect was small, and the influence on the mechanical properties of hydrate sediments was the same as for the decrease of hydrate saturation caused by decomposition. Therefore, no significant change in the strength of methane hydrate sediments was observed at this stage. Once the stress–strain relationships of the three stages were obtained, it was found that with the increase of the strain rate of triaxial compression, the stress–strain relationships of different particle sizes were gradually transformed into plastic strain. Ultimately, this became the strain yielding phenomenon, wherein the strain of the specimen could be recovered at the beginning; meanwhile, the structure of the specimen was gradually destroyed with the increasing of the axial load, and the damage became irreversible.

3.2. Strength Relationship of Sediments with Different Clay Contents

Figure 4 shows the relationship between the maximum shear strength of sediments with different clay contents and the ratio of clay to hydrate sediments from low to high. Maximum shear strength generally refers to the stress corresponding to the peak value of the stress–strain curve. However, in the case wherein the stress–strain curve has no significant peak value, the stress value at which the axial strain rate reached 15% was used. It can be seen from the figure that the maximum destructive strength of hydrate sediments was generally reduced as the proportion of clay in the sediment increased, and the maximum shear strength of the hydrate sediments of the pure sand skeleton was the highest. The reason for this may be that the strength of the material itself had a greater influence on the strength of the hydrate sediment. In general, the strength of clay is much lower than that of quartz sand. As the proportion of clay in the sediment increased, the content of quartz sand decreased correspondingly, such that the strength of the skeleton decreased. Additionally, some of the clay adhered to the surfaces of the sand grains, with some acting as a lubricant among the quartz sand particles, thereby reducing the cohesive force and internal friction, resulting in insufficient stability of the skeletal structure during the shearing process. The higher the clay content was, the more obvious this effect became.

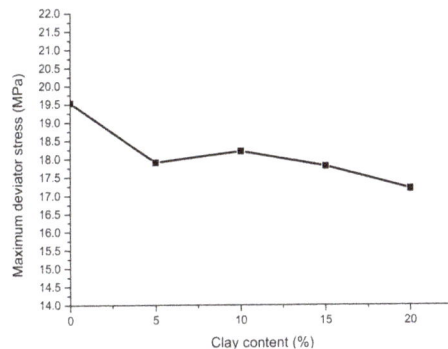

Figure 4. Maximum deviator stress by clay content (%).

The secant stiffness modulus E_{50} of sediments with different clay ratios is shown in Figure 5. This modulus is the line connecting the origin and 50% of the maximum shear strength in the stress–strain curve of the sediments. The slope of the line segment reflects the mean stiffness characteristics of the hydrate sediment. As is shown, when the hydrate content of the hydrate sediments increased under the same temperature and pressure conditions, the secant stiffness modulus gradually decreased. This reveals that as the clay content increased, the stiffness of the hydrate sediments gradually decreased.

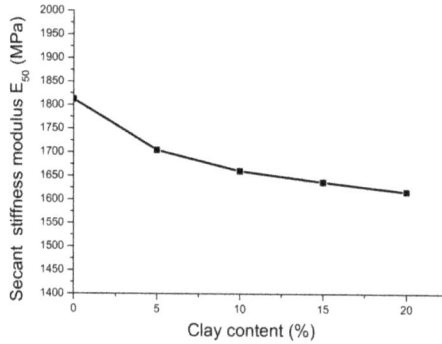

Figure 5. Secant stiffness modulus E_{50} of sediments by clay content (%).

In Figure 6, the initial tangential modulus E_0 of sediments with different clay ratios are shown. Under the same temperature and pressure conditions, the initial tangential modulus of sediments with different clay contents (as skeletons) also differed. As the clay content in the hydrate sediments increased, the initial tangential modulus of the hydrate sediments decreased gradually, while that of the pure sand (as the sediment skeleton) was the highest. Meanwhile, the initial tangential modulus of sediments with different clay contents was inconsistent. This may be due to the fact that the strength of clay is lower than that of quartz sand. The higher the clay content was in our experiments, the lower the proportion of quartz sand became. Moreover, the sediment skeletons with different clay contents were also different. The clay filled in the pores of the quartz sand skeleton, and the porosity with different clay contents also changed, which resulted in the cementation of the hydrate between the skeletons. In the same way, the initial strength of the entire hydrate sediment was decreased as the clay content increased.

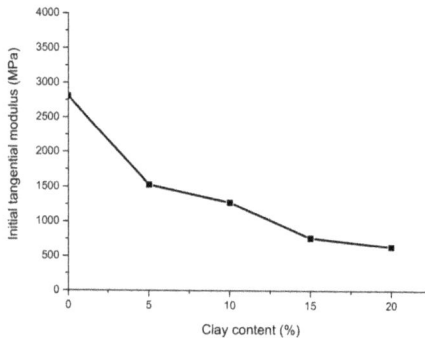

Figure 6. Initial tangential modulus of sediments by clay content (%).

3.3. Volumetric Strain Relationship of Sediments with Different Clay Contents

Figure 7 shows the volumetric strain curves for sediments with different clay proportions. It can be seen that in the first stage of shearing (i.e., when the axial strain rate was 0–2.2%), the volume of all hydrate sediment samples slightly increased as the strain rate increased and the magnitude of the change was maintained at ~0.5. The reason for this may be that in the first stage, the pore pressure of the experiment was always maintained at 8.0 MPa, which is higher than the equilibrium pressure of the phase, and the hydrate was hardly decomposed. Thus, the relationship between the hydrate and the sediment skeleton did not have much of an influence. This caused the hydrate sediment skeleton to not undergo much deformation. Additionally, the effective confining pressure at this stage was always 2.0 MPa, and such a pressure condition was not enough to destroy the particles of the hydrate sediment. Therefore, the interaction between the particles and the relative position did not change much, which resulted in there being no significant change in the volume of the entire hydrate sediment sample.

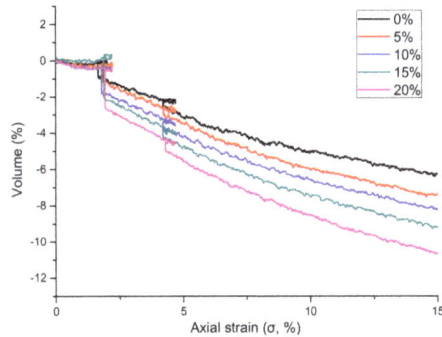

Figure 7. Volume–strain curves of sediments by clay content (%). The first stage represents axial strain rates of 0–2.2%, the middle stage accounts for rates of 2.2–4.7%, and the final stage represents strain rates of 4.7–15%. Different colors correspond to different clay contents.

In the second and third stages of the experiments, the axial strain rates were 2.2–4.7% and 4.7–15%, respectively, and the volume of hydrate sediments with different clay contents increased with the axial strain rate. With the reduction of hydrate content, an important clipping phenomenon occurred. The volumetric change in the second stage was generally >2.0%, while the shear shrinking phenomenon was more obvious in the third stage. The change in the volume of the sample with 20% clay content in the third stage even reached 5.0%. The reason for this may be that the pore pressures of the hydrate sediment samples in these two stages were lower than the phase equilibrium pressure of the hydrate at the same temperature, especially in the third stage. Thus, during the shearing process, the hydrate in the sample gradually decomposed, resulting in the formation of pores in the middle of the hydrate sediment skeleton. Under the influence of a certain effective confining pressure, some of the finer particles of clay filled the skeleton under the actions of the free water and gas generated by decomposition. Additionally, due to the large effective confining pressures in these two stages, some of the sediment particles may have been shredded. Under the external pressure, the shredded particles would have experienced larger positional changes, resulting in a change in the skeleton of the hydrate sediment sample, and ultimately in the volumetric reduction of the entire sample.

With the increase of clay content, the volumetric shearing effect of hydrate sediments became more obvious, likely because the strength of the clay was less and its compressibility was higher than that of the quartz sand. Therefore, as the clay content increased, the volume became more compressed. Figure 8 shows the grain size curve of the sediments before and after the triaxial compressive shearing of 40–60 μm specimens. It can be seen from this figure that the particle sizes before shearing were larger than afterward. This means that after shearing, the higher effective confining pressure and axial load became part of the sand. Additionally, the volume of the hydrate sediment sample suddenly declined each time the axial load pressure was unloaded during the experiment. This may have been due to the fact that these two stages comprised a process of hydrate decomposition and drainage. First, the depressurized hydrate was decomposed, and a certain amount of gas and the free water generated by decomposition were discharged, such that many pores were generated among the hydrate sediment samples. Moreover, in the process of free water and gas discharge, sand and clay also underwent substantial displacement under the movement of the free water and methane gas, which could have significantly changed the skeletal structure of hydrate sediments, especially when they were not decomposed. Finally, the sediment skeleton played a very important role in cementation, and the originally fixed skeleton was destroyed after decomposition; because of the increase in the effective confining pressure, the compression of the sample became more pronounced.

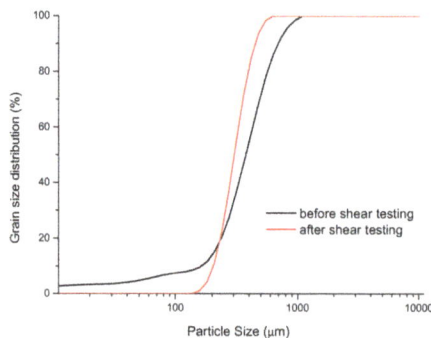

Figure 8. Grain size distribution curves of 40–60 mesh particle sizes before (black line) and after (red line) shear testing.

4. Conclusions

The stress–strain relationship of hydrate sediments containing clay (kaolin) during shearing is mainly divisible into three stages, namely elastic strain, plastic strain, and the strain yielding stage. In multi-stage triaxial shear tests with discontinuous depressurization, the mechanical strength varied with hydrate decomposition. With the increase of the axial strain rate under triaxial compression, all experiments showed that the elastic strain was gradually transformed into plastic strain. During the course of these experiments, the strain yielding phenomena changed, which indicates that the strain of the specimen could be restored at the beginning, but with the progressive loading of the axial load, the structure of the specimen was gradually destroyed, and the damage became irreversible.

The proportion of clay in the methane sediment affected the mechanical strength of the hydrate sediment sample. This may have been because the clay particles were smaller and could fill the pores of the coarser sediment skeleton, while the clay offered less support to the skeleton, and the hydrate more strongly cemented the skeleton with increasing clay content. As the proportion of clay increased, the strength of the entire sample gradually decreased. The clay also affected the volumetric strain during methane hydrate shearing. Regardless of the amount of clay added, the hydrate sediments shrank, but as the hydrates gradually decomposed, the shrinkage of the sediments became more pronounced as the clay content increased. The reason for this may be that the strength of the clay was less than that of the quartz sand, and the change in volume during the shearing process was greater than that of the quartz sand. As the hydrate decomposed, the pores between the quartz sand skeletons gradually increased, and the smaller clay particles filled the pores of the skeleton under the external shearing force, thereby causing the shearing behavior observed.

Author Contributions: D.L. (Dongliang Li), D.L. (Deqing Liang), and X.W. conceived and designed the experiments; Z.W. performed the experiments; D.L. (Dongliang Li) and Z.W. analyzed the data and wrote the paper.

Funding: This work was supported by the National Natural Science Foundation of China (grant nos. 51661165011, 51474197 and 41674076), the National Key Research & Development Plan of China (grant no. 2017YFC0307305), the Guangdong Province MEDProject (grant no. GDME-2018D002), and the Natural Science Foundation of Guangdong Province (grant no. 2018B0303110007).

Conflicts of Interest: The authors declare no conflict of interest.

References

1. Sloan, E.D. *Clathrate Hydrates of Natural Gas*, 2nd ed.; CRC Press: Boca Raton, FL, USA, 1998.

2. Klauda, J.B.; Sandler, S.I. Global distribution of methane hydrate in ocean sediment. *Energy Fuels* **2005**, *19*, 459–470. [CrossRef]

3. Collett, T.S. Energy resource potential of natural gas hydrates. *AAPG Bull.* **2002**, *86*, 1971–1992.

4. Milkov, A.V. Global estimates of hydrate-bound gas in marine sediments: How much is really out there? *Earth-Sci. Rev.* **2004**, *66*, 183–197. [CrossRef]

5. Li, G.; Li, X.S.; Tang, L.G.; Zhang, Y. Experimental investigation of production behavior of methane hydrate under ethylene glycol injection in unconsolidated sediment. *Energy Fuels* **2007**, *21*, 3388–3393. [CrossRef]

6. Tang, L.G.; Li, X.S.; Feng, Z.P.; Li, G.; Fan, S.S. Control mechanisms for gas hydrate production by depressurization in different scale hydrate reservoirs. *Energy Fuels* **2007**, *21*, 227–233. [CrossRef]

7. Komatsu, H.; Ota, M.; Smith, R.L.; Inomata, H. Review of CO_2–CH_4 clathrate hydrate replacement reaction laboratory studies – Properties and kinetics. *J. Taiwan Inst. Chem. Eng.* **2013**, *44*, 517–537. [CrossRef]

8. Kwon, T.-H.; Cho, G.-C. Submarine Slope failure primed and triggered by bottom water warming in oceanic hydrate-bearing deposits. *Energies* **2012**, *5*, 2849–2873. [CrossRef]

9. Gan, H.Y.; Wang, J.S.; Hu, G.W. Submarine Landslide Related to Natural Gas Hydrate within Benthal Deposit. *J. Seismol.* **2004**, *24*, 117–181.

10. Winters, W.; Walker, M.; Hunter, R.; Collett, T.; Boswell, R.; Rose, K.; Waite, W.; Torres, M.; Patil, S.; Dandekar, A. Physical properties of sediment from the Mount Elbert Gas Hydrate Stratigraphic Test Well, Alaska North Slope. *Mar. Pet. Geol.* **2011**, *28*, 361–380. [CrossRef]

11. Winters, W.; Waite, W.; Mason, D.; Gilbert, L.; Pecher, I.; Waite, W. Methane gas hydrate effect on sediment acoustic and strength properties. *J. Pet. Sci. Eng.* **2007**, *56*, 127–135. [CrossRef]

12. Hyodo, M.; Nakata, Y.; Yoshimoto, N.; Ebinuma, T. Basic Research on the mechanical behavior of methane hydrate-sediments mixture. *J. Jpn. Geotech. Soc. Soils Found.* **2005**, *45*, 75–85.

13. Hyodo, M.; Li, Y.; Yoneda, J.; Nakata, Y.; Yoshimoto, N.; Nishimura, A.; Song, Y. Mechanical behavior of gas-saturated methane hydrate-bearing sediments. *J. Geophys. Res. Solid Earth* **2013**, *118*, 5185–5194. [CrossRef]

14. Hyodo, M.; Wu, Y.; Nakashima, K.; Kajiyama, S.; Nakata, Y. Influence of fines content on the mechanical behavior of methane hydrate-bearing sediments. *J. Geophys. Res. Solid Earth* **2017**, *122*, 7511–7524. [CrossRef]

15. Li, Y.; Song, Y.; Yu, F.; Liu, W.; Wang, R. Effect of confining pressure on mechanical behavior of methane hydrate-bearing sediments. *Pet. Explor. Dev.* **2011**, *38*, 637–640. [CrossRef]

16. Li, Y.; Song, Y.; Liu, W.; Yu, F. Experimental research on the mechanical properties of methane hydrate-ice mixtures. *Energies* **2012**, *5*, 181–192. [CrossRef]

17. Song, Y.; Yu, F.; Li, Y.; Liu, W.; Zhao, J. Mechanical property of artificial methane hydrate under triaxial compression. *J. Nat. Gas Chem.* **2010**, *19*, 246–250. [CrossRef]

18. Song, Y.; Zhu, Y.; Liu, W.; Zhao, J.; Li, Y.; Chen, Y.; Shen, Z.; Lu, Y.; Ji, C. Experimental research on the mechanical properties of methane hydrate-bearing sediments during hydrate dissociation. *Mar. Pet. Geol.* **2014**, *51*, 70–78. [CrossRef]

19. Masui, A.; Haneda, H.; Ogata, Y.; Aoki, K. Effects of methane hydrate formation on shear strength of synthetic methane hydrate sediments. In Proceedings of the Fifteenth International Offshore and Polar Engineering Conference, Seoul, Korea, 19–24 June 2005.

20. Miyazaki, K.; Masui, A.; Sakamoto, Y.; Aoki, K.; Tenma, N.; Yamaguchi, T. Triaxial compressive properties of artificial methane-hydrate-bearing sediment. *J. Geophys. Res. Space Phys.* **2011**, *116*, 6. [CrossRef]

21. Kajiyama, S.; Wu, Y.; Hyodo, M.; Nakata, Y.; Nakashima, K.; Yoshimoto, N. Experimental investigation on the mechanical properties of methane hydrate-bearing sand formed with rounded particles. *J. Nat. Gas Sci. Eng.* **2017**, *45*, 96–107. [CrossRef]

22. Hyodo, M.; Li, Y.; Yoneda, J.; Nakata, Y.; Yoshimoto, N.; Nishimura, A. Effects of dissociation on the shear strength and deformation behavior of methane hydrate-bearing sediments. *Mar. Pet. Geol.* **2014**, *51*, 52–62. [CrossRef]

23. Jang, J.; Santamarina, J.C. Hydrate bearing clayey sediments: Formation and gas production conepts. *Mar. Pet. Geol.* **2016**, *77*, 235–246. [CrossRef]

24. Boswell, R. Is Gas Hydrate Energy Within Reach? *Science* **2009**, *325*, 957–958. [CrossRef] [PubMed]

25. Boswell, R.; Collett, T.S. Current perspectives on gas hydrate resources. *Energy Environ. Sci.* **2011**, *4*, 1206–1215. [CrossRef]

26. Dai, S.; Lee, C.; Santamarina, J.C. Formation history and physical properties of sediments from the Mount Elbert Gas Hydrate Stratigraphic Test Well, Alaska North Slope. *Mar. Pet. Geol.* **2011**, *28*, 427–438. [CrossRef]

27. Li, Y.; Hu, G.; Wu, N.; Liu, C.; Chen, Q.; Li, C. Undrained shear strength evaluation for hydrate-bearing sediment overlying strata in the Shenhu area, northern South China Sea. *Acta Oceanol. Sin.* **2019**, *38*, 114–123. [CrossRef]

28. Choi, J.H.; Dai, S.; Lin, J.S.; Seol, Y. Multi-stage triaxial tests on laboratory-formed methane hydrate-bearing sediments. *J. Geophys. Res. Solid Earth* **2018**, *123*, 3347–3357. [CrossRef]

29. Li, D.; Wu, Q.; Wang, Z.; Lu, J.; Liang, D.; Li, X. Tri-axial shear tests on hydrate-bearing sediments during hydrate dissociation with depressurization. *Energies* **2018**, *11*, 1819. [CrossRef]

30. Ghiassian, H.; Grozic, J.L. Strength behavior of methane hydrate bearing sand in undrained triaxial testing. *Mar. Pet. Geol.* **2013**, *43*, 310–319. [CrossRef]

energies

MDPI

Article

The Effects of Coupling Stiffness and Slippage of Interface Between the Wellbore and Unconsolidated Sediment on the Stability Analysis of the Wellbore Under Gas Hydrate Production

Jung-Tae Kim [1], Ah-Ram Kim [2], Gye-Chun Cho [1,*], Chul-Whan Kang [1] and Joo Yong Lee [3]

[1] Department of Civil and Environmental Engineering, Korea Advanced Institute of Science and Technology, Daejeon 34141, Korea; kimjungtae@kaist.ac.kr (J.-T.K.); jeremoss@kaist.ac.kr (C.-W.K.)
[2] Department of Infrastructure Safety Research, Korea Institute of Civil Engineering and Building Technology, Gyeonggi 10223, Korea; kimahram@kict.re.kr
[3] Petroleum and Marine Resource Devision, Korea Institute of Geoscience and Mineral Resources, Daejeon 34132, Korea; jyl@kigam.re.kr
* Correspondence: gyechun@kaist.edu

Received: 10 October 2019; Accepted: 29 October 2019; Published: 1 November 2019

Abstract: Gas hydrates have great potential as future energy resources. Several productivity and stability analyses have been conducted for the Ulleung Basin, and the depressurization method is being considered for production. Under depressurization, ground settlement occurs near the wellbore and axial stress develops. For a safe production test, it is essential to perform a stability analysis for the wellbore and hydrate-bearing sediments. In this study, the development of axial stress on the wellbore was investigated considering the coupling stiffness of the interface between the wellbore and sediment. A coupling stiffness model, which can consider both confining stress and slippage phenomena, was suggested and applied in a numerical simulation. Parametric analyses were conducted to investigate the effects of coupling stiffness and slippage on axial stress development. The results show that shear coupling stiffness has a significant effect on wellbore stability, while normal coupling stiffness has a minor effect. In addition, the maximum axial stress of the well bore has an upper limit depending on the magnitude of the confining stress, and the axial stress converges to this upper limit due to slipping at the interface. The results can be used as fundamental data for the design of wellbore under depressurization-based gas production.

Keywords: methane hydrate; shear/normal coupling stiffness; slippage at the interface; wellbore stability analysis; depressurization method

1. Introduction

Gas hydrates are solid crystalline compounds which consist of water and guest molecules [1]. Gas hydrates are formed under certain sets of high pressure and low temperature conditions, outside of which the gas and water species typically remain in separate phases [2]. The guest molecules are gas molecules such as methane, ethane, propane, or carbon dioxide. These guest molecules are combined by hydrogen-bonded water. This natural gas is a premium fuel because it burns cleanly and produces less carbon dioxide [3]. According to the latest research, approximately 230 natural gas hydrate deposits have been investigated globally, with reserves of about 1.5×10^{15} m^3 of natural gas [4]. Most natural gas hydrate deposits appears to be in the form of 'structure I', with methane as the trapped guest molecule, and its fraction is more than 90% [5,6].

Methane hydrate is also an important future energy resource for South Korea. The national projects, Ulleung Basin gas hydrate expeditions 1 and 2 (UBGH1 in 2007 and UBGH2 in 2010), were conducted to investigate the hydrate reserves and characteristics of gas hydrate-bearing sediments of the Ulleung Basin in the East sea of Korea [7–9]. Based on the data of UBGH2, the estimated amount of natural gas-hydrate deposits in Ulleung Basin ranged from 4.4×10^6 to 9.2×10^9 m^3 [10–12]. Recently, Bo et al. [13] suggested deterministic estimation from rock physics modeling and pre-stack inversion, and estimated the total gas-hydrate and gas resource volume in Ulleung Basin as about 8.43 $\times 10^8$ m^3 and 1.38×10^{11} m^3, respectively. These amounts can provide usable energy for more than thirty years to the whole nation.

During depressurization to dissociate the methane gas from HBSs, significant ground settlement occurs. This is because of the strength and stiffness reduction induced by the phase change of hydrate (e.g., the state of hydrate converts from ice crystal state to liquid or gaseous state), and the increase of effective stress, which is the stress carried by the soil. Field test for gas hydrate production has several technical problem according to the complexity of mechanism aforementioned, and needs huge budget. Thus, it is essential to perform the numerical analysis to ensure the productivity of gas hydrate and stability of production facilities before the field test. Thermal-hydraulic-mechanical (THM) coupled numerical analysis should be performed to simulate the mechanism of gas hydrate production [18]. Several numerical coupled simulators based on kinetic and equilibrium model have been developed (e.g., TOUGH+Hydrate [19], HydrateResSim [20], MH21 [21], and STOMP-HYD-KE [22]). Kim et al. [23,24] also developed the THM coupled simulator using FLAC3D, and verified with cylindrical core experimental data [25]. The input parameters (e.g., boundary condition, intrinsic hydrate reaction rate, intrinsic permeability, initial hydrate saturation, overall heat conductivity, wellbore heating temperature, bottom hole pressure, etc.) and constitutive models (e.g., permeability model, stiffness model, and heat transfer model) for numerical analysis significantly affects both the energy recovery potential and geological hazards prevention [26–28]. Kim et al. [26] provided a comprehensive estimation for model parameters and properties based on vast data from field seismic surveys in Ulleung basin and laboratory experimental results. Numerical studies on the efficiency and productivity of gas hydrate production have been carried out continuously, while stability analysis for the hydrate-bearing sediments or wellbore has not been much considered, although stability analysis is essential to field production [29–34].

As depressurization is applied in a sediment, frictional forces are evolved at the interface between the production wellbore and the soil layer due to the stiffness differences of materials. These frictional forces result in axial stresses induced on the production wellbore [35–37]. For this reason, soil–structure interaction (SSI) analysis should be conducted before the field test to properly evaluate the stability of the wellbore and HBS. The concepts of shear and normal coupling stiffness (also called interface stiffness) based on the linear Coulomb shear strength criterion are widely used to simulate interfacial stress behavior in numerical analysis [38]. The non-linear behavior of the soil–structure interface and the displacement behavior were investigated according to the interface models [39]. However, there is not much research on the stability analysis of the interface between the production wellbore and HBSs, which is related to the complex mechanism of gas hydrate production in the oceanic environment. Only a few studies have considered the wellbore stability during gas hydrate production [24]. Geological stability was assessed for vertical and horizontal well production scenarios from a displacement perspective [40]. Numerical analyses were performed to investigate the geomechanical behavior of HBS (e.g., pressure, temperature, hydrate saturation, and volumetric stain) and wellbore stability during methane production [24,41]. Kim et al. [24]

also restrictively considered the interface properties related to the interaction behavior between the sediment and wellbore. Previous studies have conducted the numerical analysis using the interface model, which considers mainly stiffness of sediments ignoring the confining stress change. The stability analysis during gas hydrate production has to consider the variation of confining stress according to depressurization. However, the research which conducted the stability analysis considering confining stress on interface model has not been published yet.

In this study, the authors investigated the effects of coupling stiffness and slippage phenomena on the stability of the wellbore under gas hydrate production. The present paper describes the concept of coupling stiffness, and the limitation of coupling stiffness model used in FLAC3D. The coupling stiffness models considering the confining stress were derived from the results of experimental tests using artificial Ulleung basin specimen, and applied to the T-H-M simulator developed in previous research [24]. Qualitative numerical analyses were performed to investigate the effects of coupling stiffness and slippage phenomena on the stability of wellbore under depressurization. More specifically, parametric analysis was conducted to investigate the trend of the development of axial stress according to the shear and normal coupling stiffness, and effects of slippage phenomena on the evolution of axial stress of wellbore. Additionally, the relationship between the development of axial stress on wellbore and geotechnical behavior of hydrate bearing sediments under depressurization was investigated.

2. Thermal–Hydraulic–Mechanical Simulation for Wellbore Stability

The mechanism of gas hydrate production from hydrate bearing sediments (HBS) is the complex reaction related to thermal, hydraulic, and mechanical (T-H-M) behaviors. This section will provide a brief description of the constitutive models for the T-H-M simulator, which was developed by Kim et al. [24]. An explanation is also provided of the limitations of the existing coupling stiffness model used in the FLAC3D software, with a suggestion for a new linear regression model derived through experimental tests. Additionally, the concept of stress evolution to consider the slippage at the interface is described.

2.1. Thermal–Hydraulic–Mechanical Coupled Simulator

Constitutive Models for T-H-M Simulator

In this study, a three-dimensional T-H-M coupled simulator, which was developed by Kim et al. [24], was used for evaluating the development of axial stress on the wellbore. The T-H-M simulator is based on the commercial finite difference method program, FLAC3D. By solving coupled thermal, hydraulic, and mechanical constitutive models, the T-H-M coupled simulator can model phase behavior, flow of fluids and heat transfer of hydrate deposit. The constitutive models used for simulated T-H-M are briefly described in this section. An elastoplastic Mohr-Columb model is used in the mechanical analysis. To consider the phase behavior, equilibrium hydrate pressure (P_e) and the corresponding temperature (T) are calculated by Kamath's equation [42]:

$$P_e = \exp\left(\alpha + \frac{\beta}{T}\right) \tag{1}$$

where P_e is the equilibrium hydrate pressure (kPa), T is the temperature corresponding to pressure (K), and the model parameters α and β are 42.047 and −9332, respectively. The rate of hydrate decomposition can be estimated [43]:

$$\frac{\partial n_g}{\partial t} = K_d A_s n S_h (P_e - P) \tag{2}$$

$$K_d = K_0 \exp\left[-\frac{\Delta E_a}{RT}\right] \tag{3}$$

where n_g is the moles of methane in the hydrate, K_d is the kinetic constant (mol m^{-2}Pa^{-1}s^{-1}), A_s is the specific surface area of the hydrate-bearing sediment (3.75×10^5 m^2), n is the porosity, S_h is the hydrate saturation, P_e is the equilibrium pressure (kPa), P is the present pressure (kPa), K_0 is the intrinsic kinetic constant (1.24×10^5 mol m^{-2}Pa^{-1}s^{-1}), ΔE_a is the activation energy, R is the gas constant (8.314 J mol^{-1}K^{-1}), and T is the temperature (K). A specific value, $(-\Delta E_a/R) = 9400 \pm 545$ K, was applied in this study.

Multi-phase flow was modeled by Darcy's law [44] and the relative permeability of gas and water were considered by van Genuchten model (1980) [45]:

$$k_r^w = S_e^b \left[1 - \left(1 - S_e^{1/a} \right)^a \right]^2 \tag{4}$$

$$k_r^g = (1 - S_e)^c \left[1 - S_e^{1/a} \right]^{2a} \tag{5}$$

where k_r^w is the relative permeability of water, k_r^g is the relative permeability of methane gas, and S_e is the effective saturation. And a, b, and c are the van Genuchten parameters. The dissociating process of hydrate is endothermic reaction. To consider the thermal reaction, the energy balance equation was used as follows:

$$c^T \frac{\partial T}{\partial t} + \nabla q^T + \rho_w c_w q_w \cdot \nabla T + \rho_g c_g q_g \cdot \nabla T - q_h^T = 0, \tag{6}$$

$$c^T = \rho_s c_s + n \left(S_h \rho_h c_h + S_g \rho_g c_g + S_w \rho_w c_w \right) \tag{7}$$

$$q_h^T = \frac{\partial n_g}{\partial t} \Delta H \tag{8}$$

where ΔT is the change in temperature per unit time (K), c^T is the effective specific heat (J/kg/K), q is the seepage-velocity vector (m/s), ρ is the density (kg/m^3), and c is the specific heat (J/kg/K), and the hydrate dissociation enthalpy change ΔH is 56.9 kJ/mol. The subscripts s, g, w, and h represent the soil, gas, water, and hydrate, respectively. More detailed overall of development and verification of T-H-M simulator had been described in Kim et al. [23,24].

2.2. Interface Model

2.2.1. Concept of Force Transfer at the Interface

During dissociation of methane hydrate from HBS by the depressurization method, ground settlement can occur due to the increase of effective stress, which is induced by decreasing the pore water pressure. At this moment, the frictional forces are generated at the interface between the production wellbore and the soil layer due to the stiffness difference of material, and draws the production wellbore. Therefore, it is essential to consider the interface characteristics for accurate stability analysis of wellbore during depressurization method. The concepts of shear and normal coupling stiffness (also called interface stiffness) have been widely used in numerical analysis to consider the interface characteristics [39,46,47]. In this study, the FLAC3D was used to estimate the wellbore stability considering the interface characteristics under the methane hydrate production. FLAC3D provides interfaces that are characterized by Coulomb sliding and/or tensile and shear bonding. The normal and shear forces at the interface are determined at calculation time $(t + \Delta t)$ through the following equations:

$$F_n^{(t+\Delta t)} = k_n u_n A + \sigma_n A \tag{9}$$

$$F_{si}^{(t+\Delta t)} = F_{si}^{(t)} + k_s \Delta u_{si}^{(t+(1/2)\Delta t)} A + \sigma_{si} A \tag{10}$$

where $F_n^{(t+\Delta t)}$, and $F_{si}^{(t+\Delta t)}$ are the normal and shear forces (N) at time $(t + \Delta t)$, respectively. u_n is the absolute normal penetration of the interface node into the target face (m), and Δu_{si} is the incremental relative shear displacement vector (m). σ_n is the additional normal stress added due to interface stress

(Pa), and σ_{si} is the additional shear stress vector due to interface stress initialization. k_n and k_s are the normal and shear stiffness (Pa/m). A is the representative area associated with the interface node (m²) [38]. The normal and shear coupling stiffness act like spring constants at the interface. The concept of load transfer at the interface considering the coupling stiffness in numerical analysis is as shown in Figure 1.

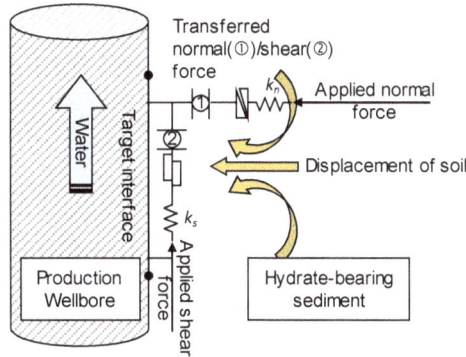

Figure 1. Concept of transferring the normal and shear stress at the interface under the depressurization method.

2.2.2. Coupling Stiffness Model in FLAC3D

The shear and normal coupling stiffness are usually determined experimentally by measuring the stress and deformation through the direct shear test or triaxial test [46,48]. In FLAC3D, the shear and normal coupling stiffness to estimate the frictional force are derived by empirical model (Equation (11)). This model is a function of the bulk and shear modulus of soil, and considers that the shear and normal coupling stiffness has equal value:

$$k_s = k_n = max\left[\frac{K + \frac{4}{3}G}{\Delta Z_{min}}\right] \qquad (11)$$

where k_s, k_n are shear coupling stiffness and normal coupling stiffness (MPa/m), respectively; K and G are bulk and shear modulus (MPa), respectively; Δz_{min} is the smallest width of an adjoining zone in the normal direction; and the max [ ] notation indicates that the maximum value over all zones adjacent to the interface is to be used [38]. While the coupling stiffness model used in FLAC3D considers only the stiffness of soil, it reveals that the normal and shear coupling stiffness are largely affected by the interface properties (i.e., confining stress, roughness, interfacial cohesion, interfacial friction angle, etc.) [49]. In particular, many studies have found that confining stress has a significant effect on the coupling stiffness through experimental tests [48,49]. Therefore, it is necessary for the interface model to consider confining stress for accurate stability analysis of the wellbore.

2.2.3. Linear Regression Models from Lab-Scale Experimental Tests

Laboratory-scale tests were performed to investigate the correlation between shear and normal coupling stiffness with confining stress. Direct shear tests, which consider the shearing interface between the wellbore and sediment, were conducted to evaluate the shear coupling stiffness. Figure 2a shows experiment set-up of direct shear test. In this experiments, artificial specimen of the Ulleung Basin core sample, which has D_{10} = 52 um, D_{30} = 90 um, D_{60} = 145 um, was used to simulatethe sediments of the pilot test site (UBGH2-6) and a STS316L disk was used as production wellbore surface. During the shearing, confining stress was maintained with measuring displacement and load. In addition, consolidation tests were conducted to determine the effect of confining stress on the normal coupling stiffness. Displacement was measured while axial stress was applied on the top of specimen. Figure 2b presents simple diagram of experiment set-up of consolidation tests on the simulated interface between artificial specimen and wellbore surface. Normal stress was applied until 650 kPa with measuring displacement of the specimen. The experiment was repeated with various specimen height from 30 mm to 70 mm.

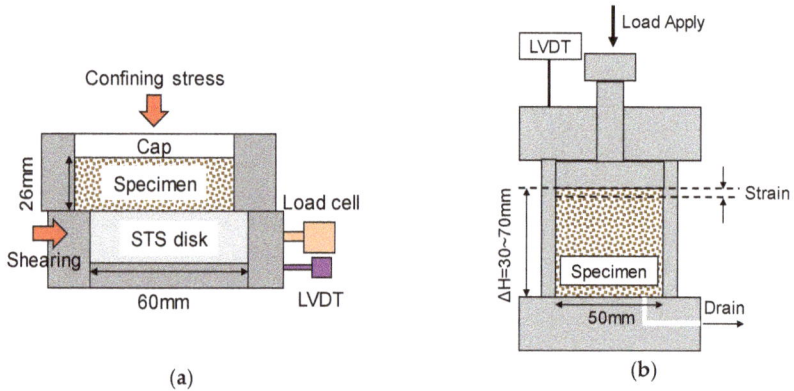

Figure 2. Schematic diagram of lab-scale tests: (**a**) direct shear test; (**b**) consolidation test.

Figure 3a shows measured data of shear stress and displacement while shearing wellbore surface and sediments. The shear coupling stiffness can be derived from the measured data from the slope of the relationship between shear stress and displacement. The slope was calculated from the peak shear stress point which represents highest stiffness level at the residual stress condition. Figure 3a shows that the shear coupling stiffness increases with the increment of confining stress.

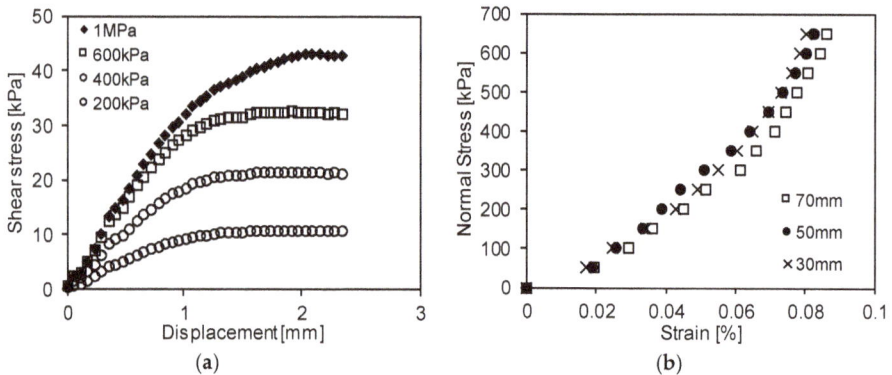

Figure 3. Results of lab-scale tests: (**a**) stress-displacement curve of the consolidation test; (**b**) stress-strain curve of the direct shear test.

Figure 3b shows the expressed stress with the strain rate instead of displacement because displacement is affected by size of the specimen. Interface stiffness is calculated as stress divided by displacement, which is derived from strain. The effective distance concept was utilized to convert strain into displacement, where the diameter of the production wellbore was considered as the effective distance.

The linear regression interface models derived from the experimental data are as shown in Figure 4. The shear and normal coupling stiffness have a linear trend with the confining stress. The models of the shear and normal coupling stiffness considering confining stress are as shown in Equations (12) and (13):

$$k_s = 45 \cdot \sigma'_c - 2.13, \tag{12}$$

$$k_n = 30 \cdot \sigma'_c + 11.9, \tag{13}$$

where σ'_c is confining stress (MPa).

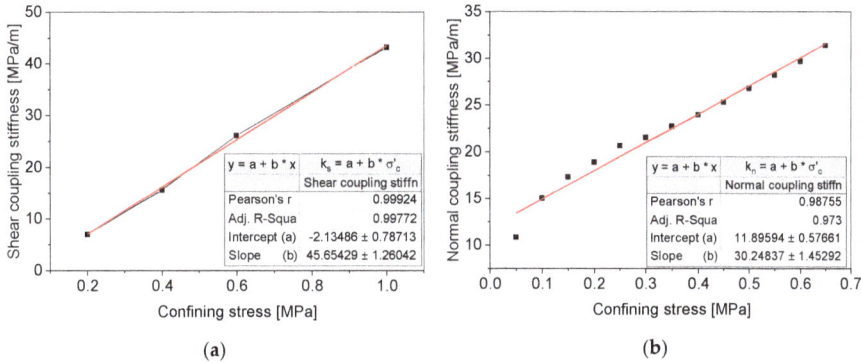

Figure 4. Linear regression models with confining stress: (**a**) shear coupling stiffness model; (**b**) normal coupling stiffness model.

The proposed models and existing model used in FLAC3D were compared. As shown in Figure 5, the existing model shows a constant value with confining stress because it is a function only for the modulus. In contrast, the proposed models show linear trends with confining stress, and estimate the shear and normal coupling stiffness differently. Through the results of the direct shear test and consolidation test, it is regarded that it is more reasonable to use the proposed models for simulating the stability analysis of interface behavior.

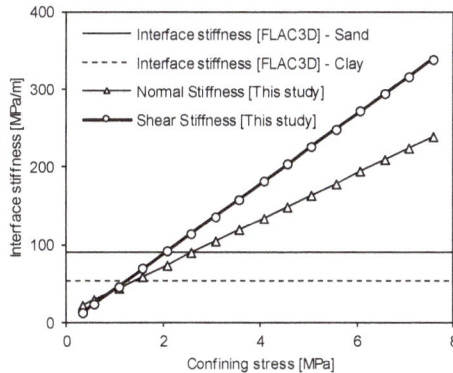

Figure 5. Comparison of interface stiffness models.

2.3. Slippage at the Interface

2.3.1. Concept of Wellbore Stress Evolution

During production of methane gas from the HBS under the depressurization method, hydrate-bearing sediment is settled with the increase of effective stress. According to Equation 10, the transferred shear stress at the interface is a function of coupling stiffness, shear deformation, shear stress vector, and skin area. Therefore, the transferred shear stress at the interface is proportional to the ground subsidence due to the shear deformation term. This leads to development of the axial stress on the wellbore due to the normal and shear coupling stiffness during ground subsidence. According to the Coulomb stress-strength criterion, the shear stress at the interface cannot exceed the shear strength of soil (Figure 6). Therefore, the shear failure at the interface occurs when the shear stress at the interface reaches the shear strength. After the shear failure of the sediments, the friction between the sediments and wellbore is constant, and there is no additional evolution of axial stress on the wellbore due to the slippage phenomenon on the interface.

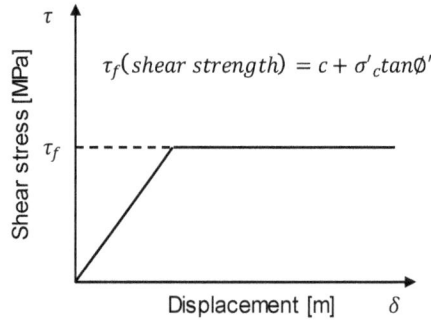

Figure 6. Comparison of interface stiffness models.

2.3.2. Maximum Axial Stress

The axial stress on the production wellbore, which is induced by the compaction of sediments, cannot be developed over the specific upper bound value according to the Coulomb stress-strength criterion. This critical stress value is defined as the maximum axial stress in this study. The maximum axial stress on the wellbore varies with confining stress during depressurization because the axial stress on the wellbore is a function of the confining stress. The axial stress is developed by the shear stress on the interface area induced by ground subsidence. The maximum axial stress is derived in the following order. At first, axial stress on the wellbore, $[\sigma_a]_t$ (*Pa*), can be estimated as follows:

$$[\sigma_a]_t = \tau \cdot A_s / A_c \tag{14}$$

where τ ($\tau = f(k_s, u_{si})$) is the shear stress (Pa), u_{si} is the displacement in shear direction (m), A_s is the skin area (m^2), and A_c is the cross-section area of wellbore (m^2). When the shear stress reaches the shear strength of sediments, slippage occurs and the axial stress at this time is the maximum axial stress. Therefore, the maximum axial stress can be expressed as:

$$[\sigma_a]_{max} = \tau_f \cdot A_s / A_c \tag{15}$$

where $[\sigma_a]_{max}$ is the maximum axial stress of the wellbore (*Pa*), and τ_f ($\tau_f = c + \sigma'_c tan\phi'$) is the shear strength in each production period (*Pa*), c is the cohesion (*Pa*), σ'_c is the confining stress (*Pa*), and ϕ' is the friction angle (°). Because the shear stress cannot exceed the shear strength of sediments, the maximum axial stress converges to a specific value of constant confining stress. However, even in the same sediment, the shear strength depends on the confining stress, because the shear strength is a function of confining stress. This means that the maximum axial stress can vary with the confining stress.

2.4. Algorithm for Stability Analysis of the Wellbore

Procedure for Estimating the Axial Stress on the Wellbore

The present paper suggests an algorithm for estimating the axial stress on the wellbore. The proposed algorithm consists of the aforementioned constitutive models and the algorithm for simulating the mechanism of the depressurization method. The flow chart of the proposed algorithm is shown in Figure 7. The detailed descriptions of each stage are as follows. At the first stage, initial shear and normal coupling stiffness are evaluated through the initial confining stress near the wellbore. Second, shear and normal coupling stiffness are updated with changes of pore pressure and effective stress by depressurization. Third, shear strengths of sediments are evaluated according to the updated parameters and maximum axial stresses are calculated by shear strength. Fourth, the maximum stress and axial stress, which is induced by the settlements of HBS, are compared. If the axial stress generated by the subsidence is larger than the maximum axial stress, then shear coupling stiffness is set to zero in order to simulate the slippage between the wellbore and sediments. Otherwise shear coupling stress remains at the same value as in the previous stage and the procedure of depressurization is continued. These procedures iterate until the analytical flow time is equal to the target time.

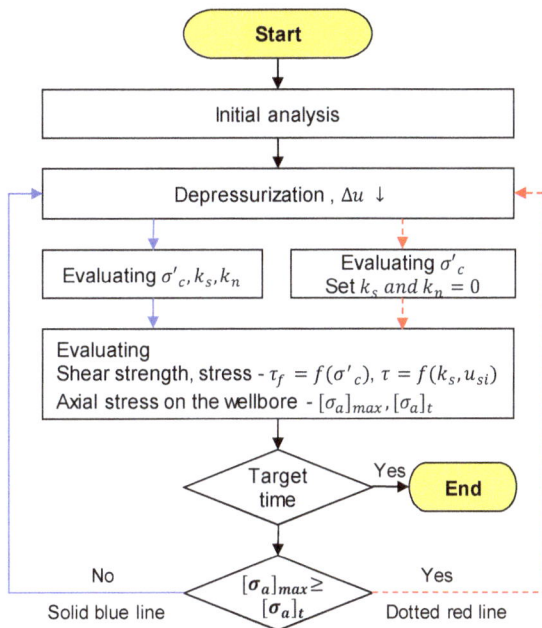

Figure 7. Algorithm for well-bore stability analysis of simulator.

2.5. Target Site and Input Parameters

Geotechnical engineers in Korea have searched for potential testbed sites for producing gas hydrate and researched mechanisms during production. UBGH2-6 is one of the sites explored during the UBGH2 project and has been established as a pilot test site. The geometry of UBGH2-6 and structure of the production well are shown in Figure 8. The HBS is located at 140 to 160 mbsf (meters below sea floor) and methane hydrate is buried in sand layers in this range. The depressurization method was selected as the production method. From the previous research, it is revealed that the productivity of gas hydrate and stability of hydrate-bearing sediments are significantly affected by bottom hole pressure (BHP), and the appropriate bottom hole pressure for the pilot test is 9 MPa for Ulleung basin [24]. For this reason, we decided UBGH2-6 as a target site for stability analysis, and depressurization method as a production method. The depressurization was conducted in the depth range from 140 to 160 mbsf with a depressurization rate of 0.5 MPa/h until the bottom hole pressure (BHP) was 9 MPa. The input parameters (i.e., properties of wellbore, mechanical, hydraulic, and thermal properties of HBS) for the numerical analysis were taken from Kim et al. [24,26], and are summarized in Tables 1 and 2.

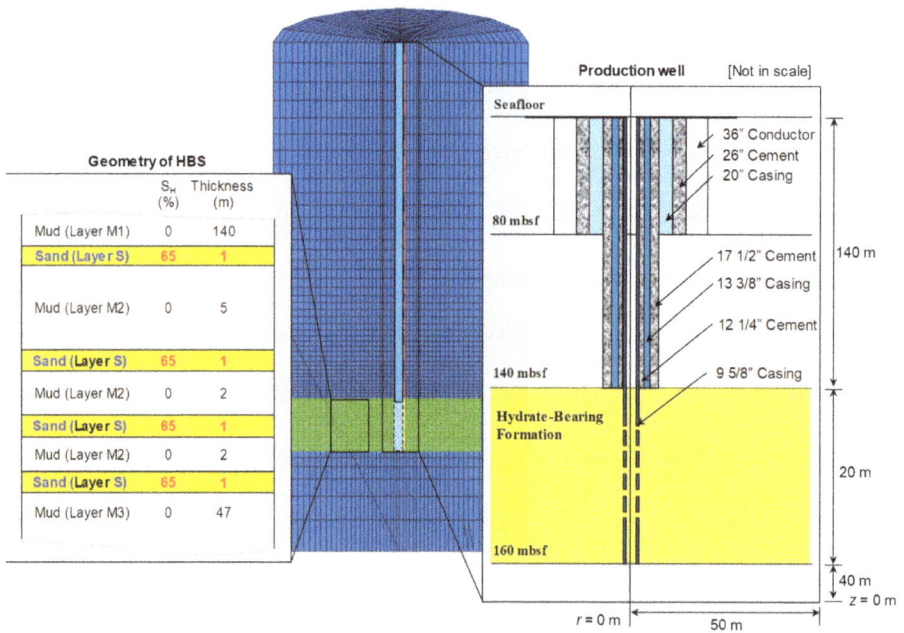

Figure 8. Schematic diagram of modeled geometry, hydrate-bearing sediment (HBS), and casing structure [24].

Table 1. Mechanical properties of the wellbore [24].

Property	Conductor	20-in. Casing	13-3/8-in. Casing	9-5/8-in. Casing	9-5/8-in. Casing Screen Part	Cement Grout
Diameter (inch)	36	20	13.375	9.625	9.625	-
Thickness (inch)	1.5	1	0.514	0.472	0.472	-
Elastic modulus (GPa)	200	200	200	200	120	3.47
Density (kg/m^3)	7897	7897	7897	7897	7897	1040
Yield strength (MPa)	390	390	758	758	454.8	1.74
Cohesion (MPa)	-	-	-	-	-	17.39
Friction angle (deg)	-	-	-	-	-	30

Table 2. Thermal, hydraulic, and mechanical input properties used in this study [26].

Category	Parameters	Value	Category-	Parameters	Value
Geologic conditions	Hydrostatic pressure at seafloor (MPa)	21.9		Bulk density of sand (Layer S; kg/m^3)	1700
-	Temperature at seafloor (°C)	0.482	Mechanical properties	Bulk density of mud (Layer M1; kg/m^3)	1500
-	Geothermal gradient (°C/km)	112		Bulk density of mud (Layer M2; kg/m^3)	1610
-	Hydrate occurrence zone (mbsf)	140–153	-	Bulk density of mud (Layer M3; kg/m^3)	1640
-	Initial hydrate saturation in sand (Layer S; %)	65	-	Methane hydrate density (kg/m^3)	910
-	Initial hydrate saturation in mud (%)	0.0	-	Young's modulus of sand (Layer S; MPa)	40
-	Salinity (wt%)	3.45	-	Young's modulus of mud (Layer M1; MPa)	15
Thermal properties	Thermal conductivity of sand (W/m K)	1.45	-	Young's modulus of mud (Layer M2; MPa)	18
-	Thermal conductivity of mud (W/m K)	1.00	-	Young's modulus of mud (Layer M3; MPa)	20
Hydraulic properties	Porosity of sand (Layer S; -)	0.45	-	Poisson's ratio of sand (Layer S; -)	0.25
-	Porosity of mud (Layer M1; -)	0.69	-	Poisson's ratio of mud (Layer M1, M2, and M3; -)	0.35
-	Porosity of mud (Layer M2; -)	0.67	-	Friction angle of sand (Layer S; deg)	25
-	Porosity of mud (Layer M3; -)	0.63	-	Friction angle of mud (Layer M1, M2, and M3; deg)	22
-	Residual water saturation, S_r^w (-)	0.1	-	Cohesion of sand (Layer S; kPa)	35
-	Residual gas saturation, S_r^g (-)	0.01	-	Cohesion of mud (Layer M1, M2; kPa)	30
-	-	-	-	Cohesion of mud (Layer M3; kPa)	40
Van Genuchten parameters	P_0 (kPa)	2.2	Properties related to the hydrate dissociation	Molecular mass of gas, M_g (g/mol)	16.042
-	a	0.6		Molecular mass of water, M_w (g/mol)	18.016
-	b	0.5		Molecular mass of hydrate, M_h (g/mol)	124.14
-	c	0.5		Hydrate number, N_h	6
-	-	-	-	Phase equilibrium model parameters, α, β	42.047, −9332

3. Results and Analysis

This section describes the results of stability analysis of HBS and the parametric study to evaluate the effectiveness of each parameter. The first part of this section shows the results of the stability analysis of HBS during gas hydrate production and describes the relationship between the geotechnical behavior and axial stress evolution on the wellbore. The second section shows the results of the parametric study regarding the effects of coupling stiffness and confining stress on the axial stress of the wellbore. The third section shows the effects of the slippage at the interface on the axial stress.

3.1. Stability Analysis of HBS During Gas Hydrate Production

Geotechnical Behaviors Near the Wellbore During Gas Production

This section describes the effects of geotechnical behavior on the development of axial stress on the wellbore. The ground subsidence occurs due to increased effective stress during the gas hydrate production from the hydrate bearing sediments (HBS). The stability analysis of HBS was performed to examine the geotechnical behavior during gas hydrate production, and to determine the relationship between the geotechnical behavior and evolution of axial stress on the wellbore. The coupling stiffness model of FLAC3D (Equation (11)) was applied in this stability analysis. As shown in Figure 9, x-axis displacements (i.e., lateral displacement) increase during depressurization. Until 12 hours after depressurization, no significant lateral displacements were observed. At 30 days after the beginning of gas production, the maximum lateral displacements of about 0.04 m occurred on both sides of wellbore at HBS.

Additionally, ground subsidence and heave occur with dissociation of hydrate as shown in Figure 10. Ground subsidence occurs from the seafloor at initial stage of depressurization. From 7 days after depressurization, ground heave occurs because of suction pressure (i.e., depressurization rate, 0.5 MPa/h) near the production well. The maximum value of subsidence occurs about 0.22 m at the seafloor, and the maximum value of heave occurs about 0.03 m at the bottom of the production well. The aforementioned maximum displacement of HBS is about 0.22 m and is 1.1% of the total depth of HBS. Despite the relatively small displacement, the large axial stress on the production wellbore can occur due to the high elastic modulus of the wellbore. From the distribution of x- and z-axis displacements, it can be deduced that the distribution of confining stress will be similar to the distribution of x- and z-axis displacements and the maximum axial stress will occur at the position where the z-axis (vertical direction) displacement is zero.

Distribution of confining stress under depressurization is shown in Figure 11. The development of the confining stress distribution with production period shows a rhomboid shape slightly shifted downward. Based on the previous results of lateral and vertical displacements, this shape can be explained. A rhomboid shape is induced by the distribution of lateral displacement, which shows maximum displacement at the middle of HBS. In addition, the reason for slightly shifting the maximum value is because of imbalance between the subsidence and ground heave (i.e., the neutral point appears slightly below from the middle). From these results, it is inferred that compressive stresses are generated on the wellbore at HBS, and the maximum axial compressive stress will occur at the point where displacement is zero (i.e., a point slightly below the middle of HBS). For reasons similar to those mentioned above, the maximum confining stress was about 11.4 MPa at slightly below the middle of HBS as shown in Figure 11h. Through the stability analysis of HBS during gas hydrate production, we can predict that the production wellbore at HBS will be subjected to axial compressive stress according to the ground behavior. In addition, it was confirmed that the coupling stiffness model considering confining stress should be applied with depth for accurate stability analysis of the wellbore.

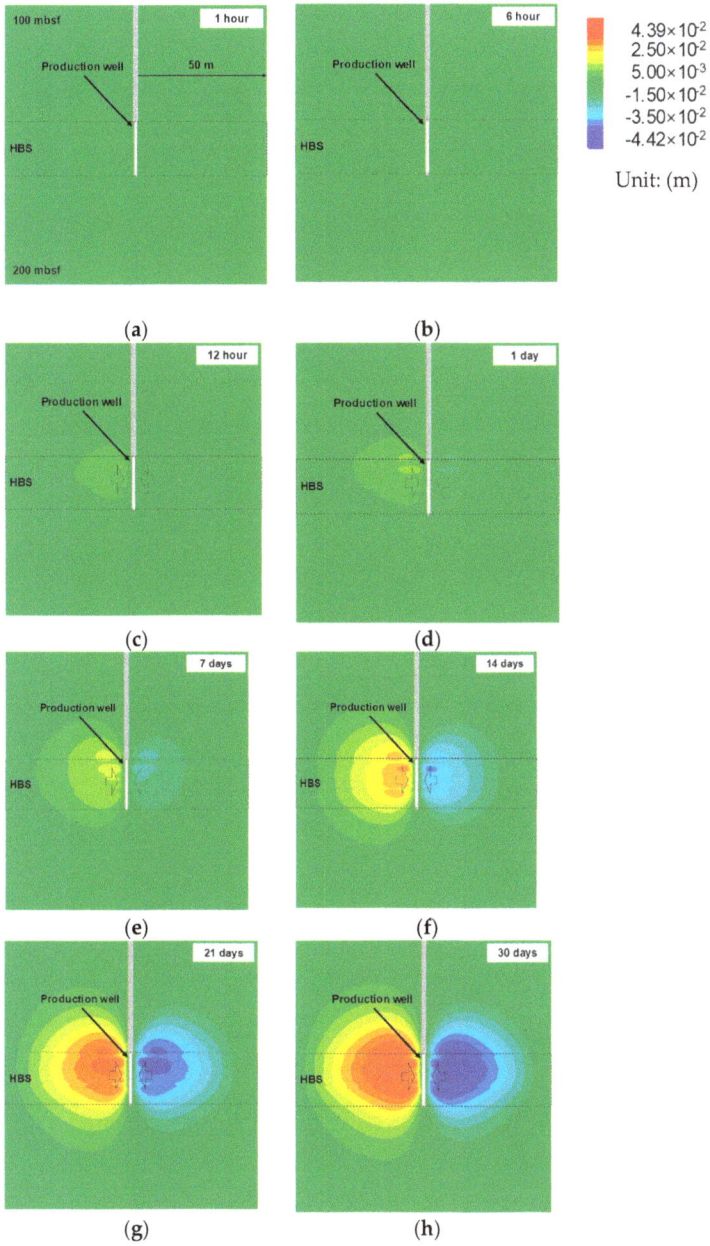

Figure 9. Distribution of x-axis displacement with production period: (**a**) after 1 hour; (**b**) after 6 hours; (**c**) after 12 hours; (**d**) after 1 day; (**e**) after 7 days; (**f**) after 14 days; (**g**) after 21 days; (**h**) after 30 days.

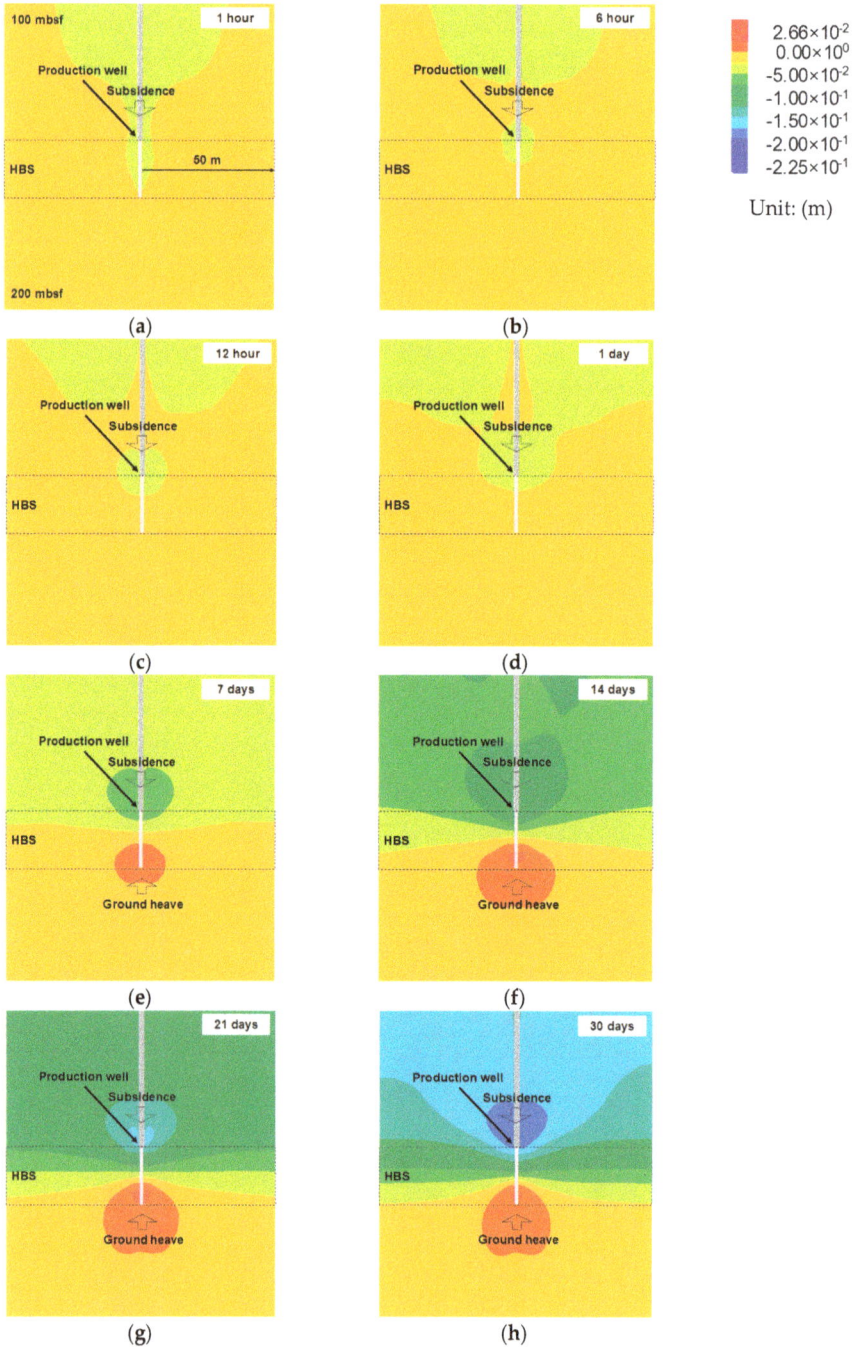

Figure 10. Distribution of z-axis displacement with production period: (**a**) after 1 hour; (**b**) after 6 hours; (**c**) after 12 hours; (**d**) after 1 day; (**e**) after 7 days; (**f**) after 14 days; (**g**) after 21 days; (**h**) after 30 days.

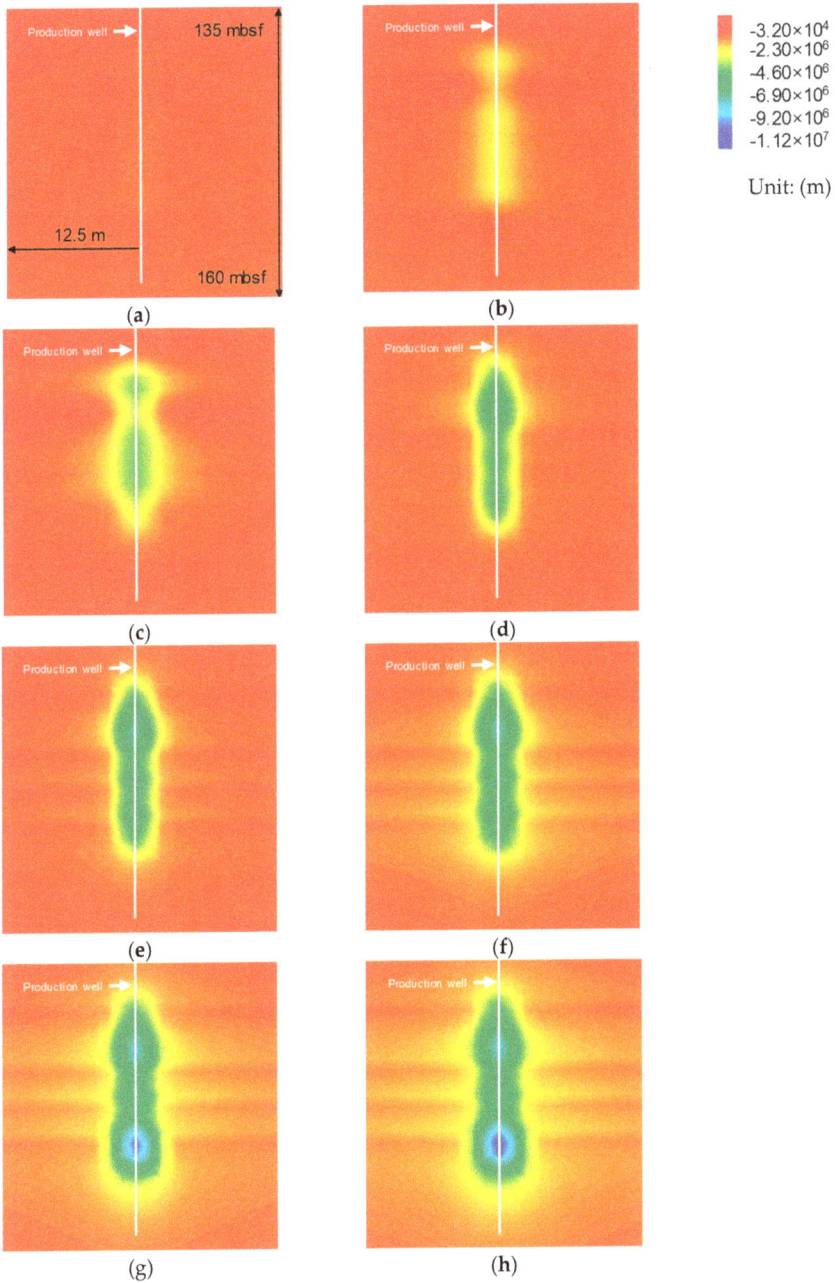

Figure 11. Distribution of confining stress with production period: (**a**) after 1 hour; (**b**) after 6 hours; (**c**) after 12 hours; (**d**) after 1 day; (**e**) after 7 days; (**f**) after 14 days; (**g**) after 21 days; (**h**) after 30 days.

3.2. Effects of Coupling Stiffness and Confining Stress on Axial Stress of Wellbore

The shear and normal coupling stiffness at the interface are widely used to simulate the interface behavior in numerical analysis [35,38,39,50]. The shear and normal coupling stiffness act as the coefficient of friction of interface between the wellbore and sediments. Typical values of the shear and normal stiffness for rock joints range from roughly 10 to 100 MPa/m for joints with soft clay in-filling [48,51,52]. According to the roughness differences between steel plate and rock, the experimental value obtained in this study is the value for the interface between soil and steel plate, which can be considered to be an appropriate value based on the aforementioned range (i.e., 10 to 100 MPa/m). Additionally, we confirmed that the shear and normal coupling stiffness vary with the confining stress (see Section 2.2.3). This section describes the effects of coupling stiffness and confining stress on axial stress of wellbore.

3.2.1. Parametric Analysis

The effective stress increases with decreasing pore pressure during depressurization. For this reason, the ground subsidence of HBS take places and induces the axial stress of the wellbore by pulling the wellbore down during production. A parametric study was performed to determine the effects of the shear and normal coupling stiffness on the stability of wellbore. Cases were defined based on the direct shear test data with confining stress of 600 kPa and consolidation test data with confining stress of 1500 kPa (case I). The effects of confining stress have not been considered in cases I to IV (using the FLAC3D model). Additionally, case V (using the new model described in Section 2.2.3, i.e., the applied coupling stiffness model considering confining stress) was defined to determine the effects of confining stress on axial stress of the wellbore. The applied cases are summarized in Table 3.

Table 3. Comparison of cases for parametric analyses.

Case	Normal Coupling Stiffness, k_n (Pa/m)	Shear Coupling Stiffness, k_s (Pa/m)	Description
I	57.41×10^6	26.11×10^6	Experimental data
II	5.74×10^6	26.11×10^6	$1/10\ k_n$ compared to case I
III	28.71×10^6	2.61×10^6	$1/2\ k_n$ and $1/10\ k_s$ compared to case I
IV	57.41×10^6	2.61×10^6	$1/10\ k_s$ compared to case I
V	$k_n = 30 \cdot \sigma'_c + 11.9$	$k_s = 45 \cdot \sigma'_c - 2.13$	Consider confining stress

3.2.2. Results of Parametric Study

The results of the parametric study are shown in Figure 12. The production period for parametric analyses is 14 days. The distribution of axial stress on the wellbore for case I is shown in Figure 12a. As shown in Figure 12a, the compressive and tensile stresses on the wellbore are developed due to the ground subsidence. The distribution of axial stress on the wellbore in other cases also showed a similar trend to case I. The maximum compressive and tensile stress of each case are shown in Figure 12b. The yield strength of the production wellbore (9 5/8″ casing) is 454.8 MPa in this study. The compressive axial stress of case I, II and V (Figure 12b) exceed the yield strength of the wellbore located at the screen parts. By contrast, results of case III and IV show compressive stresses of the wellbore lower than the yield strength and the tensile stresses are similar for all cases.

The effect of normal coupling stiffness on axial stress can be confirmed through the comparison between case I and II or case III and IV. Cases I has a normal stiffness 10 times higher than those of case II. The results show that the compressive and tensile stress increased about 1.2% (i.e., from 739 to 748 MPa) and 0.3% (i.e., from 335 to 336 MPa), respectively, while normal stress increased 10 times. On the other hand, case III has a normal stiffness twice as high as those of case IV, but show almost no difference in axial stresses. The effect of shear coupling stiffness can be confirmed by the comparison between case I and IV. These cases have 10 times difference in the shear coupling stiffness, and show that the maximum compressive stress significantly increased about 174% (i.e., from 273 to 748 MPa). From the above comparisons, it can be concluded that the shear coupling stiffness significantly affects the development of axial stress of the wellbore, while the normal coupling stiffness has little effect.

In addition, the effect of the confining stress on the axial stress development of the wellbore can be explained in case V. The maximum axial compressive stress, 1194.6 MPa, of case V was about 2.6 times greater than the yield strength of the wellbore and significantly greater than those of other cases. The considerably large development of the maximum axial compressive stress compared to other cases is due to the large shear and coupling stiffness according to the increase of confining stress. During the 14-day period of gas production, the pore pressure decreased to 9 MPa and consequently the confining stress increased until about 7.6 MPa. After 14 days of gas production, relatively large shear and normal coupling stiffness (319.8 and 239.4 MPa, respectively) were derived by the coupling stiffness models considering the confining stress. This large shear and normal coupling stiffness increased axial stress on the wellbore. From this result, it is concluded that the consideration of confining stress largely affects the development of axial stress on the wellbore. Therefore, the variation of confining stress should be considered in coupling stiffness models for accurate analysis.

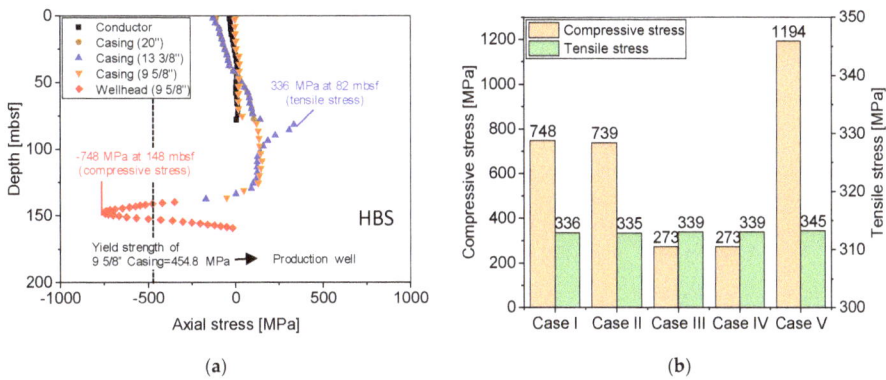

Figure 12. Distribution of axial stresses with depth (non-consideration of the slippage at the interface): (**a**) Distribution of axial stress with depth for case I; (**b**) comparison of the maximum compressive and tensile stress.

3.3. Effects of Slippage at the Interface

As described in Section 3.1, the confining stress increases during gas hydrate production. Relatively high maximum compressive stress was derived in Section 3.2 due to non-application of slippage phenomena. In this section, stability analysis was performed by considering the slippage phenomenon according to the Coulomb stress–strain criterion (see Section 2.3). Development of compressive stress with the production period is shown in Figure 13a. The pore pressure (i.e., bottom hole pressure) near the screen part decreased from the initial state, 23.5 MPa, to target pressure, 9 MPa. The depressurization rate was 0.5 MPa/h in this study. After 29 hours of gas production, the bottom hole pressure was reduced to 9 MPa and this pressure spreads out of the production wellbore. Under the influence of changes in pore pressure, the confining stress also rapidly increased in the initial state. The compressive stress sharply increased in the depressurization period and converged to a certain value (334.9 MPa for this study). The shape of the development of axial compressive stress of the wellbore was similar to the Coulomb stress-strain curve (Figure 6).

Distributions of axial stress with depth of the wellbore under depressurization are shown in Figure 13b. The maximum axial compressive stress converged to 334.9 MPa from 14 days after the beginning of depressurization. The generated maximum axial stress is about 74% of the yield strength of the wellbore, 454.9 MPa. This means that the production well will be stable until 30 days after the start of production. The maximum compressive stress converges to a constant value, and the convergence range gradually spreads to the whole range of the production wellbore. This is because the coupling stiffness applied differently with the depth of wellbore, and is considered to be zero after the slippage phenomena (see Section 2.4). Similar to the geotechnical behaviors of the previous section, the maximum axial compressive stress has been observed from about 153 mbsf in which the neutral point is described in Section 3.1.

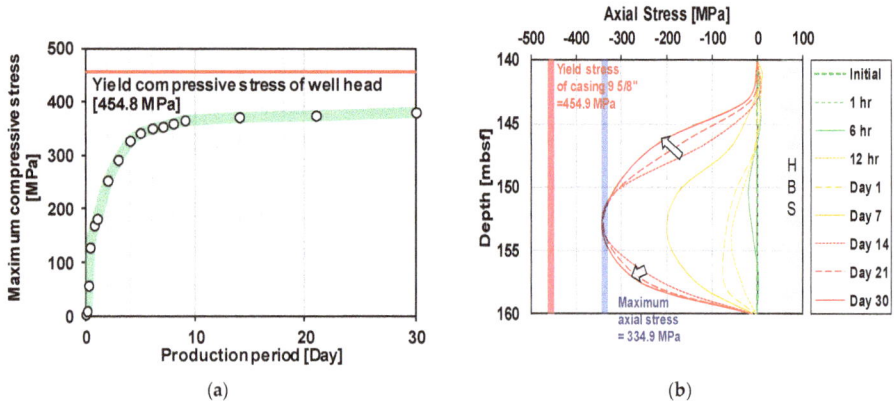

Figure 13. Results of stability analysis considering the coupling stiffness models: (**a**) Development of the maximum axial stresses with production period; (**b**) distribution of axial stress with depth.

4. Discussion

4.1. Comparison of the Models

This section suggests a suitable method for stability analysis of the wellbore based on the results of the present study. The axial stress on the wellbore was largely affected by the coupling stiffness and, as shown in the distribution of confining stress (see Figure 11), the confining stress varies with the depth of the wellbore. From the results of the direct shear test, it was revealed that the coupling stiffness changes with confining stress. However, the existing model of the shear and normal coupling stiffness in FLAC3D is the function of the shear and bulk modulus without consideration of the confining stress. Therefore, it is difficult to obtain accurate stability analysis results using the existing model in FLAC3D. This study suggests that the stability analysis for the wellbore during gas hydrate production has to consider the coupling stiffness differently with confining stress according to the depth. Parametric studies have been conducted to investigate the effect of coupling stiffness change according to the depth on the development of the axial stress on the wellbore to verify the proposed stability analysis method.

The previous section described the effects of coupling stiffness and geotechnical behavior on development of axial stress. In this section, parametric analysis was carried out to understand the effects of coupling stiffness model and slippage at the interface. Assigned cases for the parametric analysis are summarized in Table 4.

Table 4. Comparison of cases for parametric analysis.

Case	Coupling Stiffness Model	Slippage at Interface	Production Period
A	w/o consideration	w/o consideration	14 days
B	With consideration FLAC3D (constant) $k_s = k_n = \max\left[\frac{K+\frac{4}{3}G}{\Delta Z_{min}}\right]$	w/o consideration	14 days
C	With consideration the confinement dependent model (this study) $k_s = 45 \cdot \sigma'_c - 21.4$	w/o consideration	14 days
D	$k_n = 30 \cdot \sigma'_c + 11.9$	with consideration	14 days

As a control case, Case A did not apply the coupling stiffness model and the slippage at the interface. Case B applied the existing FLAC3D model, which is the function of the shear and bulk modulus without consideration of confining stress, as a coupling stiffness model. Case C applied the linear regression models derived from this study, and does not consider the slippage at the interface. Case D applied the linear regression model and also considered the slippage at the interface. The production period was set to be 14 days for all cases.

4.2. Comparison of Results According to Model Application

Results according to model application are shown in Figure 14a and comparison of the maximum compressive and tensile stresses of each case are shown in Figure 14b. Case A, which does not consider the coupling stiffness, showed almost zero axial stress on the wellbore because the transferred stress from the sediments to the wellbore was too small. Cases B and C yielded maximum axial compressive stresses of 520.4 MPa and 1194.6 MPa, respectively. This difference of the maximum compressive stress is due to the difference of coupling stiffness. The shear coupling stiffness of case C (339 MPa/m) is about 3.72 times larger than that of Case B (91 MPa/m for sand sediment) at the completion of depressurization (see Figure 5). As a result, the maximum compressive stresses of C is 2.3 times higher than those of case B. This result tends to be similar to the results of parametric studies to derive the effects of coupling stiffness (see Section 3.2). The maximum compressive stresses of both Cases B and C exceeded the yield strength of 9 5/8" casing, 454.8 MPa. Unlike the results of Cases B and C, Case D shows that the maximum axial compressive stress, 334.9 MPa, of the wellbore was lower than the yield strength of the 9 5/8" casing.

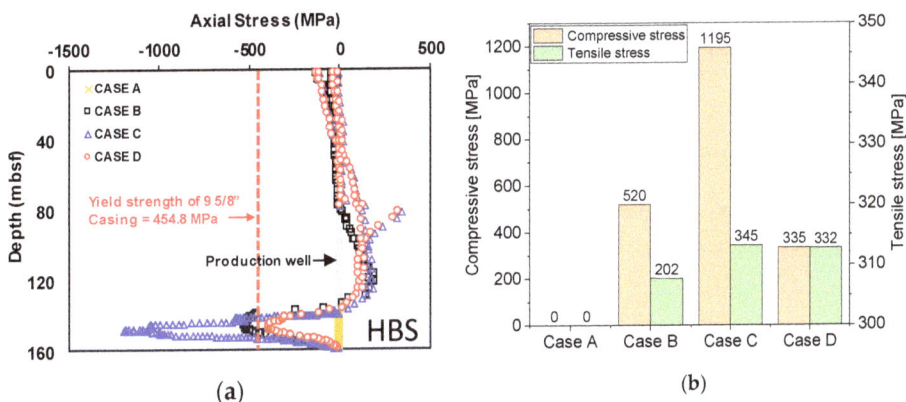

Figure 14. Results of parametric analyses: (**a**) Distribution of axial stress; (**b**) the maximum axial stresses.

Because the coupling stiffness acts as a spring constant at the interface between the wellbore and soil sediments in the numerical analysis, the large coupling stiffness induces the large axial stress according to the ground subsidence. Therefore, it is deduced that the case for non-consideration of the slippage phenomena over-estimates axial stress. From the results of parametric study, it is revealed that the proposed method, which considers the coupling stiffness differently with depth, can accurately simulate the stability of the wellbore during the gas hydrate production.

5. Conclusions

The aim of the present study was to evaluate the stability of a production wellbore under the depressurization method for gas hydrate production from hydrate bearing sediments. In order to evaluate the stability of the wellbore, it is essential to consider the interface behavior between the wellbore and hydrate bearing sediments. In this paper, an algorithm for wellbore stability analysis was suggested. The effects of the shear and normal coupling stiffness were investigated and coupling stiffness models, which considered confining stress and slippage phenomena, were suggested and applied to the algorithm. The key findings from this study are as follows:

- The shear and normal coupling stiffness have to be considered to simulate the interface between the wellbore and the ground. From the parametric analysis relating coupling stiffness to wellbore stability, shear coupling stiffness has a significant effect on the development of axial stress of the wellbore while normal coupling stiffness does not affect the development of axial stress on the wellbore.
- The shear and normal coupling stiffness are the function of confining stress. This study derived a coupling stiffness model considering confining stress by performing the direct shear test and consolidation test.
- The shear coupling stiffness has to be considered differently from the depth of the wellbore to estimate the actual development of axial stress on the wellbore.
- The compressive stress is induced at the wellbore due to the subsidence and heave of the ground.
- As the effective stress with depressurization for gas hydrate production increases, slippage occurs between the wellbore and the ground because of shear failure. After shear failure, additional axial stress on the wellbore is not developed. For this reason, the maximum generated axial stress converges to a whole range during gas hydrate production.
- Preferentially, the maximum axial stress occurs at the neutral point where displacement is zero, and gradually converges to the whole range of the wellbore. The final conclusion based on the key findings is that the coupling stiffness has to be considered differently from the depth of the wellbore, and the slippage phenomena also has to be considered to performed accurate stability analysis.

This study contains limitation and the authors suggest further work. In this study, the coupling stiffness models are valid only for Ulleung basin because the models were derived from the experimental results using Ulleung basin sediment. In order to improve the applicability of the models, it is necessary to develop them including factors (e.g., bulk and shear modulus) that can consider the characteristics of the soil type.

Author Contributions: Conceptualization, G.-C.C. and J.-T.K.; numerical analysis, J.-T.K. and A.-R.K.; experiment, C.-W.K.; analysis of data, J.-T.K. and A.-R.K.; writing—original draft preparation, J.-T.K.; writing—review and editing, A.-R.K. and G.-C.C.; supervision, G.-C.C.; project administration, J.Y.L.

Funding: This research was supported by the Ministry of Trade, Industry, and Energy (MOTIE) through the Project "Gas Hydrate Exploration and Production Study (18-1143)" under the management of the Gas Hydrate Research and Development Organization (GHDO) of Korea and the Korea Institute of Geoscience and Mineral Resources (KIGAM) and supported by the National Research Foundation of Korea (NRF) grant funded by the Korea government (MSIT) (No. 2018R1C1B6008417).

Conflicts of Interest: The authors declare no conflict of interest.

References

1. Yin, Z.; Khurana, M.; Tan, H.K.; Linga, P. A review of gas hydrate growth kinetic models. *Chem. Eng. J.* **2018**, *342*, 9–29. [CrossRef]
2. Chong, Z.R.; Yang, S.H.B.; Babu, P.; Linga, P.; Li, X.-S. Review of natural gas hydrates as an energy resource: Prospects and challenges. *Appl. Energy* **2016**, *162*, 1633–1652. [CrossRef]
3. Sloan, E.D. Fundamental principles and applications of natural gas hydrates. *Nature* **2003**, *426*, 353. [CrossRef] [PubMed]
4. Rossi, F.; Gambelli, A.M.; Sharma, D.K.; Castellani, B.; Nicolini, A.; Castaldi, M.J. Experiments on methane hydrates formation in seabed deposits and gas recovery adopting carbon dioxide replacement strategies. *Appl. Therm. Eng.* **2019**, *148*, 371–381. [CrossRef]
5. Lu, H.; Seo, Y.-T.; Lee, J.-W.; Moudrakovski, I.; Ripmeester, J.A.; Chapman, N.R.; Coffin, R.B.; Gardner, G.; Pohlman, J. Complex gas hydrate from the Cascadia margin. *Nature* **2007**, *445*, 303. [CrossRef]

6. Makogon, Y.F. Natural gas hydrates—A promising source of energy. *J. Nat. Gas. Sci. Eng.* **2010**, *2*, 49–59. [CrossRef]

7. Park, K.-P.; Bahk, J.-J.; Kwon, Y.; Kim, G.Y.; Riedel, M.; Holland, M.; Schultheiss, P.; Rose, K. Korean national Program expedition confirms rich gas hydrate deposit in the Ulleung Basin, East Sea. *Fire Ice Methane Hydrate Newsl.* **2008**, *8*, 6–9.

8. Kim, G.Y.; Yi, B.Y.; Yoo, D.G.; Ryu, B.J.; Riedel, M. Evidence of gas hydrate from downhole logging data in the Ulleung Basin, East Sea. *Mar. Pet. Geol.* **2011**, *28*, 1979–1985. [CrossRef]

9. Ryu, B.-J.; Collett, T.S.; Riedel, M.; Kim, G.Y.; Chun, J.-H.; Bahk, J.-J.; Lee, J.Y.; Kim, J.-H.; Yoo, D.-G. Scientific results of the second gas hydrate drilling expedition in the Ulleung basin (UBGH2). *Mar. Pet. Geol.* **2013**, *47*, 1–20. [CrossRef]

10. Kang, N.; Yoo, D.; Yi, B.; Bahk, J.; Ryu, B. Resources Assessment of Gas Hydrate in the Ulleung Basin, Offshore Korea. In Proceedings of the AGU Fall Meeting Abstracts, San Francisco, CA, USA, 3–7 December 2012.

11. Lee, G.H.; Bo, Y.Y.; Yoo, D.G.; Ryu, B.J.; Kim, H.J. Estimation of the gas-hydrate resource volume in a small area of the Ulleung Basin, East Sea using seismic inversion and multi-attribute transform techniques. *Mar. Pet. Geol.* **2013**, *47*, 291–302. [CrossRef]

12. Riedel, M.; Bahk, J.-J.; Kim, H.-S.; Scholz, N.; Yoo, D.; Kim, W.-S.; Ryu, B.-J.; Lee, S. Seismic facies analyses as aid in regional gas hydrate assessments. Part-II: Prediction of reservoir properties, gas hydrate petroleum system analysis, and Monte Carlo simulation. *Mar. Pet. Geol.* **2013**, *47*, 269–290. [CrossRef]

13. Bo, Y.Y.; Lee, G.H.; Kang, N.K.; Yoo, D.G.; Lee, J.Y. Deterministic estimation of gas-hydrate resource volume in a small area of the Ulleung Basin, East Sea (Japan Sea) from rock physics modeling and pre-stack inversion. *Mar. Pet. Geol.* **2018**, *92*, 597–608.

14. Holder, G.; Kamath, V.; Godbole, S. The potential of natural gas hydrates as an energy resource. *Annu. Rev. Energy* **1984**, *9*, 427–445. [CrossRef]

15. Collett, T.; Bahk, J.-J.; Baker, R.; Boswell, R.; Divins, D.; Frye, M.; Goldberg, D.; Husebø, J.; Koh, C.; Malone, M. Methane Hydrates in Nature Current Knowledge and Challenges. *J. Chem. Eng. Data* **2014**, *60*, 319–329. [CrossRef]

16. Dallimore, S.; Collett, T. Summary and implications of the Mallik 2002 gas hydrate production research well program. *Sci. Results Mallik* **2005**, *585*, 1–36.

17. Yamamoto, K.; Terao, Y.; Fujii, T.; Ikawa, T.; Seki, M.; Matsuzawa, M.; Kanno, T. Operational overview of the first offshore production test of methane hydrates in the Eastern Nankai Trough. In Proceedings of the Offshore Technology Conference, Houston, TX, USA, 5–8 May 2014.

18. Moridis, G.J.; Collett, T.S.; Boswell, R.; Hancock, S.; Rutqvist, J.; Santamarina, C.; Kneafsey, T.; Reagan, M.T.; Pooladi-Darvish, M.; Kowalsky, M. Gas hydrates as a potential energy source: State of knowledge and challenges. In *Advanced Biofuels and Bioproducts*; Springer: Berlin, Germany, 2013; pp. 977–1033.

19. Moridis, G.J. *TOUGH+ HYDRATE v1. 2 User's Manual: A Code for the Simulation of System Behavior in Hydrate-Bearing Geologic Media*; University of California: Berkeley, CA, USA, 2012.

20. Moridis, G.; Kowalsky, M.; Pruess, K. *HydrateResSim Users Manual: A Numerical Simulator for Modeling the Behavior of Hydrates in Geologic Media*; Lawrence Berkeley National Laboratory: Berkeley, CA, USA, 2005.

21. Kurihara, M.; Ouchi, H.; Masuda, Y.; Narita, H.; Okada, Y. *Assessment of Gas Productivity of Natural Methane Hydrates Using MH21 Reservoir Simulator*; Natural Gas Hydrates/Energy Resource Potential Associated Geologic Hazards: Vancouver, BC, Canada, 2004.

22. White, M. Impact of kinetics on the injectivity of liquid CO_2 into Arctic hydrates. In Proceedings of the OTC Arctic Offshore Technology Conference, Houston, TX, USA, 7 February 2011.

23. Kim, A. THM Coupled Numerical Analysis of Gas Production from Methane Hydrate Deposits in the Ulleung Basin in Korea. Ph.D. Thesis, Korea Advanced Institute of Science and Technology (KAIST), Daejeon, Korea, 2016.

24. Kim, A.-R.; Kim, J.-T.; Cho, G.-C.; Lee, J.Y. Methane Production From Marine Gas Hydrate Deposits in Korea: Thermal-Hydraulic-Mechanical Simulation on Production Wellbore Stability. *J. Geophys. Res. Solid Earth* **2018**, *123*, 9555–9569. [CrossRef]

25. Masuda, Y. Modeling and experimental studies on dissociation of methane gas hydrates in Berea sandstone cores. In Proceedings of the Third International Gas Hydrate Conference, Salt Lake City, UT, USA, 18–22 July 1999.

26. Kim, A.-R.; Kim, H.-S.; Cho, G.-C.; Lee, J.Y. Estimation of model parameters and properties for numerical simulation on geomechanical stability of gas hydrate production in the Ulleung Basin, East Sea, Korea. *Quat. Int.* **2017**, *459*, 55–68. [CrossRef]

27. Zheng, R.; Li, S.; Li, Q.; Li, X. Study on the relations between controlling mechanisms and dissociation front of gas hydrate reservoirs. *Appl. Energy* **2018**, *215*, 405–415. [CrossRef]

28. Zheng, R.; Li, S.; Li, X. Sensitivity analysis of hydrate dissociation front conditioned to depressurization and wellbore heating. *Mar. Pet. Geol.* **2018**, *91*, 631–638. [CrossRef]

29. Su, Z.; Cao, Y.; Wu, N.; He, Y. Numerical analysis on gas production efficiency from hydrate deposits by thermal stimulation: Application to the Shenhu Area, south China sea. *Energies* **2011**, *4*, 294–313. [CrossRef]

30. Ruan, X.; Song, Y.; Zhao, J.; Liang, H.; Yang, M.; Li, Y. Numerical simulation of methane production from hydrates induced by different depressurizing approaches. *Energies* **2012**, *5*, 438–458. [CrossRef]

31. Sun, Z.; Xin, Y.; Sun, Q.; Ma, R.; Zhang, J.; Lv, S.; Cai, M.; Wang, H. Numerical simulation of the depressurization process of a natural gas hydrate reservoir: An attempt at optimization of field operational factors with multiple wells in a real 3D geological model. *Energies* **2016**, *9*, 714. [CrossRef]

32. Wang, Y.; Feng, J.-C.; Li, X.-S.; Zhang, Y.; Li, G. Evaluation of gas production from marine hydrate deposits at the GMGS2-Site 8, Pearl river Mouth Basin, South China Sea. *Energies* **2016**, *9*, 222. [CrossRef]

33. Ruan, X.; Li, X.-S.; Xu, C.-G. Numerical Investigation of the Production Behavior of Methane Hydrates under Depressurization Conditions Combined with Well-Wall Heating. *Energies* **2017**, *10*, 161. [CrossRef]

34. Feng, Y.; Chen, L.; Suzuki, A.; Kogawa, T.; Okajima, J.; Komiya, A.; Maruyama, S. Numerical analysis of gas production from layered methane hydrate reservoirs by depressurization. *Energy* **2019**, *166*, 1106–1119. [CrossRef]

35. Jeong, S.; Lee, J.; Lee, C.J. Slip effect at the pile-soil interface on dragload. *Comput. Geotech.* **2004**, *31*, 115–126. [CrossRef]

36. Chen, R.; Zhou, W.; Chen, Y. Influences of soil consolidation and pile load on the development of negative skin friction of a pile. *Comput. Geotech.* **2009**, *36*, 1265–1271. [CrossRef]

37. Abdrabbo, F.M.; Ali, N.A. Behaviour of single pile in consolidating soil. *Alex. Eng. J.* **2015**, *54*, 481–495. [CrossRef]

38. Itasca, F.D. *Fast lagrangian analysis of continua in 3 dimensions*; Itasca Consulting Group Inc.: Minneapolis, MN, USA, 2005.

39. Häggblad, B.; Nordgren, G. Modelling nonlinear soil-structure interaction using interface elements, elastic-plastic soil elements and absorbing infinite elements. *Comput. Struct.* **1987**, *26*, 307–324. [CrossRef]

40. Kim, J.; Moridis, G.J.; Rutqvist, J. Coupled flow and geomechanical analysis for gas production in the Prudhoe Bay Unit L-106 well Unit C gas hydrate deposit in Alaska. *J. Pet. Sci. Eng.* **2012**, *92*, 143–157. [CrossRef]

41. Rutqvist, J.; Moridis, G.; Grover, T.; Silpngarmlert, S.; Collett, T.; Holdich, S. Coupled multiphase fluid flow and wellbore stability analysis associated with gas production from oceanic hydrate-bearing sediments. *J. Pet. Sci. Eng.* **2012**, *92*, 65–81. [CrossRef]

42. Kamath, V.A. *Study of Heat Transfer Characteristics During Dissociation of Gas Hydrates in Porous Media*; Pittsburgh University: Pittsburgh, PA, USA, 1984.

43. Clarke, M.; Bishnoi, P.R. Determination of the activation energy and intrinsic rate constant of methane gas hydrate decomposition. *Can. J. Chem. Eng.* **2001**, *79*, 143–147. [CrossRef]

44. Darcy, H.P.G. *Les Fontaines Publiques de la Ville de Dijon. Exposition et Application des Principes à Suivre et des Formules à Employer dans les Questions de Distribution d'eau*; Dalamont, V., Ed.; Libraire des Corps Imperiaux des Ponts et Chaussees et des Mines: Paris, France, 1856.

45. Van Genuchten, M.T. A closed-form equation for predicting the hydraulic conductivity of unsaturated soils 1. *Soil Sci. Soc. Am. J.* **1980**, *44*, 892–898. [CrossRef]

46. Zeghal, M.; Edil, T.B. Soil structure interaction analysis: Modeling the interface. *Can. Geotech. J.* **2002**, *39*, 620–628. [CrossRef]

47. Salemi, A.; Esmaeili, M.; Sereshki, F. Normal and shear resistance of longitudinal contact surfaces of segmental tunnel linings. *Int. J. Rock Mech. Min. Sci.* **2015**, *77*, 328–338. [CrossRef]

48. Rosso, R.S. A comparison of joint stiffness measurements in direct shear, triaxial compression, and In Situ. *Int. J. Rock Mech. Min. Sci. Geomech. Abstr.* **1976**, *13*, 167–172. [CrossRef]

49. Li, W.; Bai, J.; Cheng, J.; Peng, S.; Liu, H. Determination of coal–rock interface strength by laboratory direct shear tests under constant normal load. *Int. J. Rock Mech. Min. Sci.* **2015**, *77*, 60–67. [CrossRef]

50. Jeong, S.; Cho, J. Proposed nonlinear 3-D analytical method for piled raft foundations. *Comput. Geotech.* **2014**, *59*, 112–126. [CrossRef]

51. Kulhawy, F.H. Stress deformation properties of rock and rock discontinuities. *Eng. Geol.* **1975**, *9*, 327–350. [CrossRef]

52. Bandis, S.C.; Lumsden, A.C.; Barton, N.R. Fundamentals of rock joint deformation. *Int. J. Rock Mech. Min. Sci. Geomech. Abstr.* **1983**, *20*, 249–268. [CrossRef]

Article

Mobilized Mohr-Coulomb and Hoek-Brown Strength Parameters during Failure of Granite in Alxa Area in China for High-Level Radioactive Waste Disposal

Cheng Cheng [1],*, Nengxiong Xu [1] and Bo Zheng [2]

[1] School of Engineering and Technology, China University of Geosciences (Beijing), Beijing 100083, China; xunengxiong@cugb.edu.cn

[2] Key Laboratory of Shale Gas and Geoengineering, Institute of Geology and Geophysics, Chinese Academy of Sciences, Beijing 100029, China; zhengbo@mail.iggcas.ac.cn

* Correspondence: chengc@cugb.edu.cn; Tel.: +86-10-8232-2627

Received: 30 September 2019; Accepted: 30 October 2019; Published: 6 November 2019

Abstract: Strength parameters of the host rock is of paramount importance for modelling the behaviors of underground disposal repository of high-level radioactive waste (HLW). Mobilization of strength parameters should be studied for a better understanding and modelling on the mechanical behaviors of the surrounding rock, considering the effect of temperature induced by the nuclear waste. The granite samples cored from NRG01 borehole in Alxa candidate area in China for HLW disposal are treated by different temperatures (T = 20 °C, 100 °C and 200 °C), and then are used to carry out a series of uniaxial and tri-axial compression experiments under various confining pressures ($\sigma_3 = 0, 5, 10, 20,$ and 30 MPa) in this study. With the recorded axial stress—axial strain and axial stress—lateral strain curves, mobilization of both Mohr-Coulomb and Hoek-Brown strength parameters are analyzed with the increasing plastic shear strain. It has been found that NRG01 granite samples show generally similar cohesion weakening and friction strengthening behaviors, as well as the non-simultaneous mobilization of Hoek-Brown strength parameters (m_b and s), under the effect of various treatment temperatures. Furthermore, the samples treated by higher temperatures show lower initial values of cohesion, but their initial friction angle and m_b values are relatively higher. This should be mainly owing to the thermally induced cracks in the samples. This study should be helpful for a better modelling on the mechanical behaviors of NRG01 granite samples as the host rock of a possible HLW disposal repository.

Keywords: granite; HLW disposal; plastic strain; temperature; CWFS; damage process; yield condition; strength criterion

1. Introduction

Granite is considered as one of the most important types of host rock for geological disposal of high-level radioactive waste (HLW) [1–9]. An appropriate modelling on the mechanical behavior of granite is of great importance for site selection and design of the repository [9–13]. Specially, it should be noted that the heat induced by the nuclear waste may have considerable influences on the mechanical behavior of the host rock, so the thermal effect cannot be ignored [8,12,14–17].

There have been extensive studies on the mechanical behaviors of the host rock for HLW disposal [1,9,18–22]. In many studies, simultaneously mobilized Mohr-Coulomb strength parameters (cohesion and friction angle) were used in the modelling of the underground excavation [23–27]. In order to describe the plastic strain softening behavior of the rock, they assumed that both cohesion and friction angle degrade from the initial value to the residual value with the increasing plastic strain, and piecewise linear models were usually adopted [23–27]. However, based on a series of theoretical analyses and laboratory experimental studies on cohesive soil, granite, marble, limestone, etc., it has been found that the geo-materials always show non-simultaneous mobilization of strength components, i.e., cohesion will be degraded and friction angle will be enhanced with the increasing damage or plastic strain during the failure process [26,28–35]. This is owing to the development of cracks inside the rock decreases the cohesive strength, while the induced crack surfaces make the frictional strength increases [26,29,36]. Accordingly, the cohesion weakening friction strengthening (CWFS) model was proposed, and this model with linear equations was used for modelling the failure process of URL Mine-by tunnel [26,29,36]. A comparison study shows that CWFS model can capture the failure extent and depth of failure (DOF) of this circular excavation better than the other widely used models such as elastic model, elastic-perfectly plastic model, elastic-brittle model, etc. [26]. Thereafter, more linear CWFS models are used in the researches and give reasonable simulations on the stability of underground openings, pillars, as well as the process of crack propagation [30,37–39]. Nonetheless, it was pointed out that the linear CWFS model may result in a problematic behavior of the stress – strain curves, and a fitted non-linear CWFS model with smooth curves was proposed, which was proved to be able to capture the gradual damage process better [31]. More recently, reference [40] proposed the guidelines for the parameters selection for CWFS modelling analysis of excavations. Up to date, the concept of CWFS analyses has widely been accepted in modelling the failure of brittle rocks.

However, the studies on the mobilization of strength parameters mainly focused on the Mohr-Coulomb criterion. As an actual fact, Hoek-Brown criterion is also widely used in the modelling of field rock engineering [41,42]. There were some studies using piecewise linear models with simultaneous mobilization of Hoek-Brown strength parameters (*m* and *s*) to analyze rock behaviors [23,24,26,43]. Nevertheless, according to the above-mentioned analyses on the non-simultaneous mobilization of cohesion and friction angle, we should notice whether the Hoek-Brown strength parameters may also be mobilized non-simultaneously during the damage and failure process of rock. If the answer is yes, what is the characteristics of this mobilization? What is the relationship between the mobilized Hoek-Brown strength parameters and the rock damage or plastic strain? This has not been investigated in the published researches, and it is required to conduct a detailed study.

In the recent several decades, extensive field and laboratory researches have been carried out in the site selection of HLW disposal in China [1,2,4,10,15,20,22,44,45]. Alxa candidate area in Inner Mongolia is one of the three candidate areas with large volume of granitic rock. Figure 1 presents the location of Alxa area with two sub-areas (TMS and NRG), as well as the main geological structures around this area. More detailed information about Alxa area has been provided in reference [1]. Field investigations on the outcrops have been conducted and four boreholes (named as TMS01, TMS02, NRG01 and NRG02) with the depth of 600 m have been drilled. Laboratory experiments on the cored samples have also been carried out for studying the mechanical properties of the rock. These researches show that the granite around NRG01 borehole shows the best rock mass quality in Alxa candidate area [1]. Nonetheless, further studies should still be conducted on NRG01 granite samples with coarse grains for a better modelling on the mechanical behaviors. What is the characteristics of the mobilization of cohesion and friction angle for NRG01 granite samples? How will the heat produced by the nuclear waste affect the mechanical behavior of NRG01 granite samples during the damage and failure process? Will the mobilization of Hoek-Brown strength parameters occur for NRG01 granite samples in a simultaneous or non-simultaneous way? Is there any suitable equations to describe this mobilization? What is the mechanism?

Based on a series of systematic uniaxial and tri-axial compression experiments on NRG01 granite samples treated by different temperatures, mobilization of both Mohr-Coulomb and Hoek-Brown strength parameters have been analyzed in details. This paper is organized as follows: In Section 2, the physical and mechanical properties of the samples, the experimental setup and methods will be introduced. The experimental results will be presented in Section 3. Section 4 will provide the systematic data analyses and discussions on both the mobilization of Mohr-Coulomb and Hoek-Brown strength parameters during the failure process of NRG01 granite samples under different heat treatments. Based on the above-mentioned analyses and discussions, some conclusions will be drawn in Section 5.

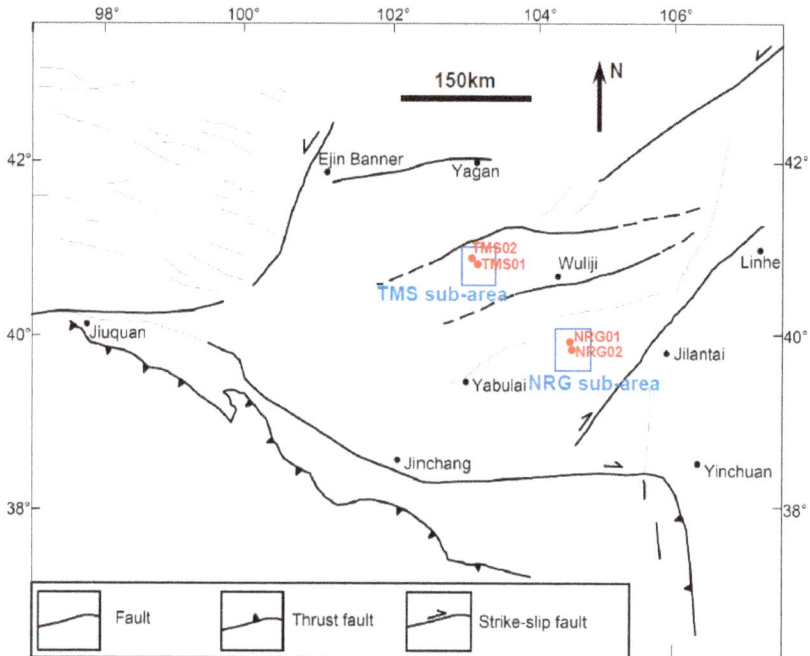

Figure 1. Schematic map of main geological structures around Alxa area. Modified after [46]. TMS01, TMS02, NRG01 and NRG02 are four boreholes drilled in TMS and NRG sub-areas.

2. Samples and Experimental Methods

NRG01 granite samples treated by different temperatures are used to conduct a series of uniaxial and tri-axial compression tests with various confining pressures. The obtained stress-strain data will be used for analyzing the mobilization of strength components during the brittle failure of granite considering the thermal effect.

2.1. Samples

The granite samples are cored from NRG01 borehole, which is one of the four 600 m-deep boreholes in Alxa area. According to the field investigations on the corresponding outcrops, RQD analyses on the drilling cores, as well as the mechanical experiments on the cored specimens in laboratory, NRG01 samples show the best structural and strength quality and thus are selected to be used for further studies [1].

The pink samples are cored from the depth of 500–600 m. The typical samples are presented in Figure 2a,b. It can be found that the samples are heterogeneous and have coarse particles. According to the observation on thin sections under polarized microscopy, the mineral contents and the grain sizes of NRG01 granite samples are analyzed and listed in Table 1 [47]. Based on the mineral components, the samples should be named as biotite syenogranite. Nevertheless, they are still called as granite samples in this paper for simplicity. Figure 2c shows a comparison on the strength values of different granite samples under various confining pressures. Apparently, NRG01 samples have higher strength than TMS01 granite samples cored from TMS01 borehole in TMS sub-area (shown in Figure 1) of Alxa candidate area. Compared with BS06 granite samples cored from Beishan candidate area in Gansu Province [48], NRG01 samples show a little lower strength under lower confinements (σ_3 = 0–10 MPa), while a little higher strength under higher confinements (σ_3 > 10 MPa). Based on peak strength values fitted with linear Mohr-Coulomb criterion, NRG01 granite samples have the cohesion of 20.1 MPa, and internal friction angle of 57.5°. Hoek-Brown criterion is also used to analyze the data, and the non-linear Hoek-Brown fitting curve is shown in Figure 2c.

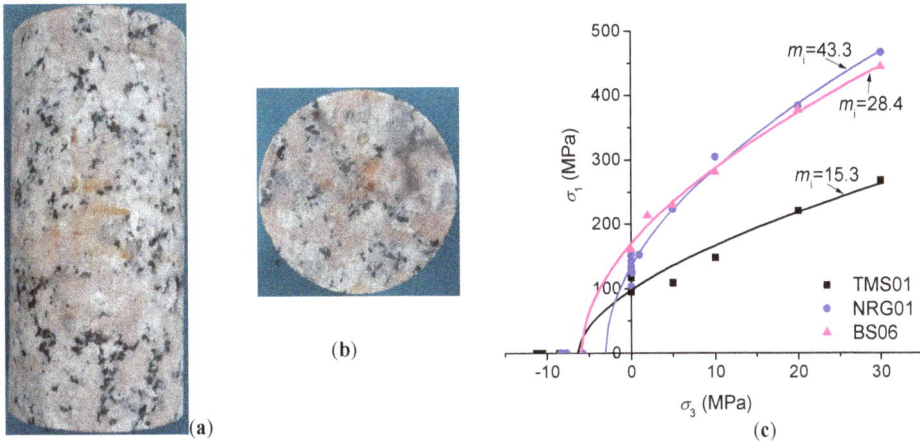

Figure 2. (**a,b**) Typical NRG01 granite samples (Height: 100mm; Diameter: 50mm) [1] and (**c**) strength of NRG01 granite under various confinements comparing with TMS01 granite as well as BS06 granite from Beishan area, Gansu Province [48].

Table 1. Mineral contents and grain sizes of NRG01 granite samples (based on [47]).

Minerals	Contents	Grain Sizes (mm)
alkali feldspar	45%	2.0–8.0
plagioclase	18%	1.3–3.0
quartz	25%	1.5–4.0
biotite	12%	0.8–1.5

2.2. Experimental Methods

A series of cylindrical NRG01 granite samples are well prepared (listed in Table 2) for uniaxial and tri-axial compression experiments under various confining pressures (σ_3 = 0, 5, 10, 20, and 30 MPa). Concerning the heat produced by the high-level radioactive waste during the long-term disposal period, the effect of temperature should also be considered in this study. According to an extensive review on the conceptual design of repositories [7,49–52], the temperature applied on the host rock will be no higher than 100 °C–120 °C. Consequently, this study focuses on the range from 20 °C (room temperature) to 200 °C.

The specimens are firstly heated in a heating cabinet to the designed temperatures as shown in Table 2. The heating rate is set as 2 °C/min. When the target temperatures are reached, the heat treated samples are used for a series of uniaxial and tri-axial compression experiments with the TAW2000 servo-control tri-axial compression test system in Key Laboratory of Shale Gas and Geoengineeirng, Chinese Academy of Sciences. It should be noted that the rock specimens cannot remain their treatment temperatures as there is not a heating system during the compression tests. The confining pressures are applied to the target values as presented in Table 2, followed by the axial loading at a constant strain rate of $1.0 \times 10^{-5} \cdot s^{-1}$. During each test, the axial and lateral strain are both measured with a set of extensometers, and the axial stress is obtained according to the axial load monitored by a force sensor. Consequently, the axial stress – axial strain curve and axial stress—lateral strain curve can be recorded for each test, and the failure characteristics of the specimens will also be observed after the experiments are completed.

Table 2. Design for the tests under different confinements and heat treatment.

NO.	Length (mm)	Diameter (mm)	Density (g/mm^3)	Confinement (MPa)	Temperature (°C)
N1-20	100.17	49.55	2.64	0	20
N1-14	100.13	49.51	2.65	5	20
N1-29	99.67	49.99	2.63	10	20
N1-7	100.31	49.99	2.65	20	20
N1-23	100.09	49.46	2.64	5	20
N1-77	100.39	50.17	2.63	0	100
N1-83	100.32	50.03	2.63	5	100
N1-85	100.15	50.02	2.65	10	100
N1-88	100.28	50.14	2.64	20	100
N1-92	100.45	50.11	2.64	30	100
N1-79	100.49	50.21	2.65	0	200
N1-82	100.32	49.74	2.64	5	200
N1-87	100.37	50.13	2.63	10	200
N1-89	100.45	49.72	2.64	20	200
N1-93	98.79	50.14	2.65	30	200

3. Experimental Results

The differential stress—axial strain curves and differential stress—lateral strain curves for all the tests are presented in Figure 3a–e. It should be noted that differential stress ($\sigma_1 - \sigma_3$) is used in these curves in order for a more consistent observation. The peak strength values and the strength envelopes fitted with Hoek-Brown criterion are shown in Figure 3f. Based on these test results, some features can be observed as follows:

(1) For the NRG01 granite samples treated by different temperatures (T = 20 °C, 100 °C and 200 °C), the differential stress – axial strain curves show the similar brittle – ductile transition behaviors with the increasing confining pressures (σ_3 = 0–30 MPa);

(2) According to the experimental results, the heat treatment by temperatures no higher than 200 °C does not have very significant influence on the stress – strain curves of the NRG01 granite samples under various confining pressures (σ_3 = 0–30 MPa). However, if we make a more careful observation, it can be found that the samples treated by higher temperature show relatively more ductile behavior during the post-peak stage;

(3) The peak strength values are also very close for the samples treated by different temperatures. This means that the heat treatment by temperatures no higher than 200 °C does not have very obvious influence on the strength values of the NRG01 granite samples under various confining pressures (σ_3 = 0–30 MPa).

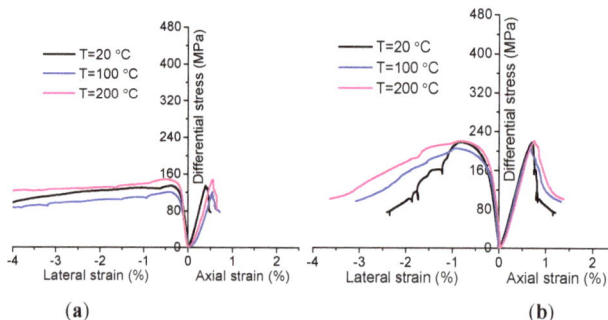

Figure 3. *Cont.*

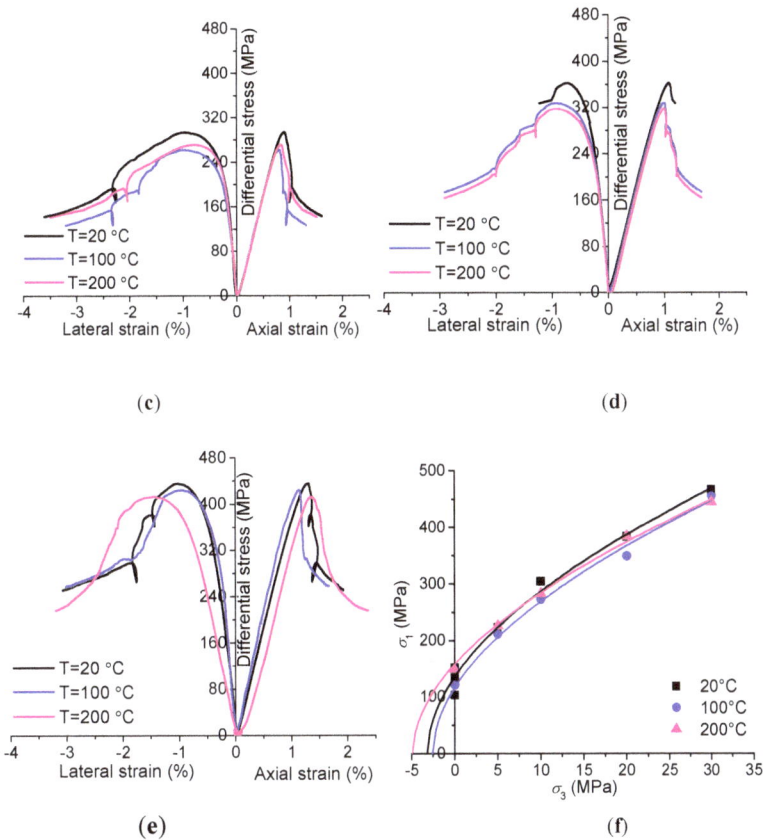

Figure 3. Differential stress—axial strain and differential stress—lateral strain curves of NRG01 granite samples treated by various temperatures under the confinement of (**a**) 0 MPa, (**b**) 5 MPa, (**c**) 10 MPa, (**d**) 20 MPa, (**e**) 30 MPa; and (**f**) the fitted peak strength envelops.

4. Data Analyses and Discussion

4.1. Mobilized Mohr-Coulomb Strength Parameters During Failure of NRG01 Granite

4.1.1. Analytical Method

According to the previous studies [26,28,31,33,53], the Mohr-Coulomb strength parameters (cohesion c and inner frictional angle φ) of rock should be mobilized dependent on rock damage or plastic parameters of geo-materials. The most widely accepted plastic parameter is the plastic shear strain γ^P, which can be obtained as the difference between the maximum and minimum principal plastic strains (ε_1^P and ε_3^P, respectively) [27,30,33,54]:

$$\gamma^P = \varepsilon_1^P - \varepsilon_3^P, \tag{1}$$

There are usually two methods to obtain the plastic strain values. One method is to differentiate the recoverable and irrecoverable strain by taking cyclic loading-unloading experiments. The plastic strain can be obtained from the irrecoverable strain in each cycle of the tests directly, however, it is quite complicated to control this type of experiment, and the data is limited by the numbers of cycles [27,33,54,55]. Therefore, another method is developed based on the assumption that the unloading curve in each cycle has the same modulus as the initial deformation modulus. In this way, a series of plastic strains can be obtained with a series of assumed loading-unloading cycles by just carrying out conventional uniaxial and tri-axial compression experiments. This method has been widely accepted and used in many studies [27,33,54] and is also employed here in this research. Figure 4 gives a sketch to illustrate this method for determining the plastic axial and lateral strains, as well as the corresponding axial stress values. A series of lines parallel with the tangent lines at the linear elastic stage of the σ_1-ε_1 curves are drawn to determine the plastic axial strain $\varepsilon_{1,i}^P$ and plastic lateral strain $\varepsilon_{3,i}^P$, respectively. The symbol i here is a series of positive integers, showing that a series of plastic strain values can be collected with this method. It should be noted that there is a gap $|OA|$ owing to the crack closure stage of σ_1-$_1$ curves, so this gap should be removed for determining the plastic axial strain $\varepsilon_{1,i}^P$:

$$\varepsilon_{1,i}^P = |OB| - |OA|, \quad \varepsilon_{1,i+1}^P = |OC| - |OA|, \tag{2}$$

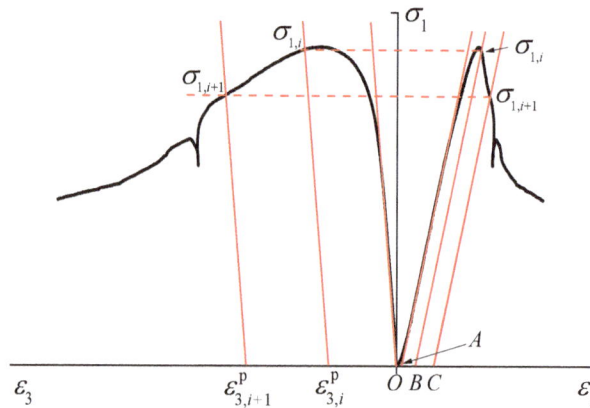

Figure 4. Sketch of the method for determining plastic strains and the corresponding stress values.

Then the plastic shear strain can be obtained as:

$$\gamma_i^P = \varepsilon_{1,i}^P - \varepsilon_{3,i}^P, \tag{3}$$

For each plastic shear strain, the corresponding maximum principal stress σ_1 is collected under different confining pressures σ_3. Thereafter, cohesion c and internal friction angle φ can be calculated by drawing Mohr circles or by linear fitting of σ_1 and σ_3 with the following equation:

$$\sigma_1 = \frac{2c\cos\varphi}{1 - \sin\varphi} + \frac{1 + \sin\varphi}{1 - \sin\varphi}\sigma_3 \tag{4}$$

The values of c and φ can then be plotted with the increasing plastic shear strain. As the temperature induced by the high-level radioactive waste may affect the mechanical behavior of the host rock, the evolutionary characteristics of c and φ are also studied for NRG01 granite under different heat treatment (20 °C–200 °C).

4.1.2. Data Analyses and Discussion

Based on the stress-strain curves of NRG01 granite under different heat treatment and confinements presented in Figure 3, as well as the methodology demonstrated in Section 4.1.1, a series of axial stress at different plastic shear strains can be plotted in Figure 5. For each plastic shear strain, a set of axial stress values under various confining pressures can be obtained to calculate the cohesion and friction angle. These values are shown in Figure 5.

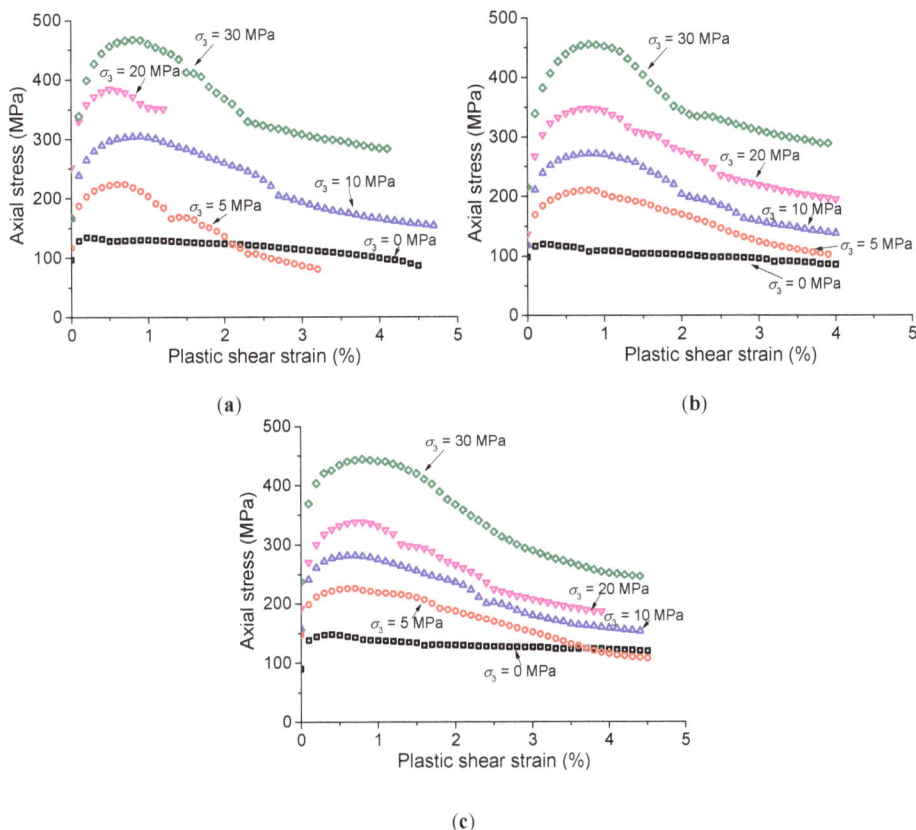

Figure 5. Evolution of maximum principal stresses of NRG01 granite samples treated by different temperatures: (**a**) T = 20 °C; (**b**) T = 100 °C and (**c**) T = 200 °C.

The results obtained in Figure 6 show that NRG01 granite samples have the generally cohesion weakening and friction strengthening (CWFS) behaviors for various treatment temperatures (room temperature to 200 °C). According to the published references [26,29,31,36], the cohesion component should be weakened to the residual value before the full mobilization of friction angle, however, it is not true for NRG01 granite treated by different temperatures. It is shown that cohesion is weakened in a gradual manner with increasing plastic shear strain, nevertheless, the friction angle increases to the peak value more immediately. As an actual fact, the test results similar to this study can also be found in references [30,31,33]. This difference has also been discussed in [29,36], and it is believed that the plastic strain limit at which the cohesion reaches the residual value or the friction angle is fully mobilized is dependent on many factors such as the rock type, grain size, heterogeneity, as well as the hoop effect owing to the cylindrical shape of the specimens, etc. More systematic studies should be carried out to learn more clearly about the exact influencing factors and the mechanism.

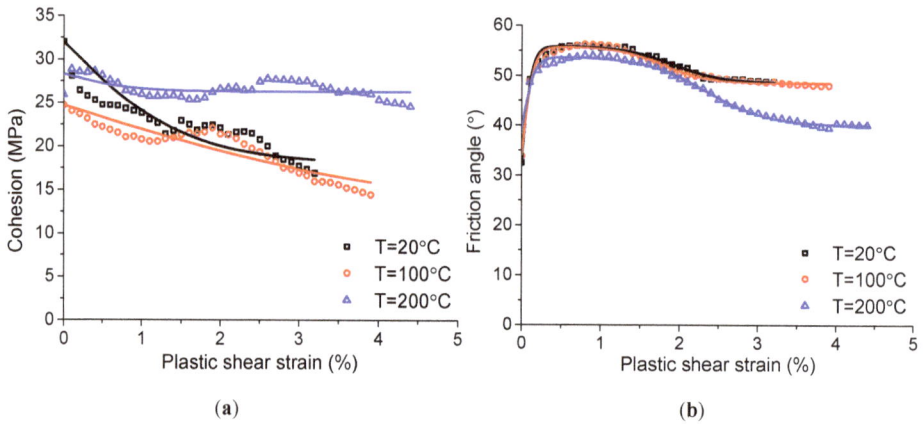

Figure 6. Mobilized (**a**) cohesion and (**b**) friction angle of NRG01 granite samples treated by different temperatures.

Based on the characteristics of the mobilized cohesion and friction angle presented in Figure 6, the generally used linear CWFS model [30,37–39] may not be suitable for NRG01 granite. For a better description of the rock behaviors, a non-linear model should be used. With the fitting Equations (5) and (6) proposed in reference [31], the mobilized cohesion and friction angle values can be well fitted as shown in Figure 6. The fitted coefficients are listed in Table 3.

$$c = c_r + (c_i - c_r)\left[2 - \frac{2}{1 + \exp\left(-5\frac{\gamma^P}{\gamma_{c,r}^P}\right)}\right],$$ (5)

where, c_i and c_r are the initial and residual values of cohesion, respectively. $\gamma_{c,r}^P$ is the plastic shear strain when the cohesion is close to the residual value.

$$\varphi = \varphi_i + (\varphi_{max} - \varphi_i)\left[\frac{2}{1 + \exp\left(-5\frac{\gamma^P}{\gamma_{\varphi,max}^P}\right)} - 1\right] - (\varphi_{max} - \varphi_r)\left[\frac{1}{1 + \exp\left(-5\frac{2\gamma^P - \gamma_{\varphi,r}^P}{\gamma_{\varphi,r}^P}\right)}\right],$$ (6)

where, φ_i is the initial value of friction angle, while φ_{max} is the maximum value, and φ_r is the residual value. $\gamma_{\varphi,max}^P$ is the plastic shear strain when the friction angle is close to its peak value, while $\gamma_{\varphi,r}^P$ is the plastic shear strain when the friction angle is close to its residual value.

Table 3. The fitted coefficients determining the mobilized Mohr-Coulomb strength parameters during the failure of NRG01 granite samples under different treatment temperatures.

Temperature (°C)	c_i	c_r	$\gamma_{c,r}^P$	R^2	φ_i	φ_{max}	φ_r	$\gamma_{\varphi,max}^P$	$\gamma_{\varphi,r}^P$	R^2
20	31.99	18.00	3.93	0.6696	32.78	55.91	48.58	0.31	3.52	0.9864
100	24.67	12.00	11.35	0.8341	34.14	55.74	48.43	0.35	3.23	0.9746
200	28.31	26.26	2.08	0.2001	39.21	53.70	39.89	0.38	4.3	0.9931

According to Figure 6, it can also be observed that the different treated temperatures may lead to a few different evolutionary behaviors of cohesion and friction angle for NRG01 granite samples. The more obvious influences are shown for the behaviors of cohesion component, i.e., the sample under room temperature (T = 20 °C) shows an apparently higher initial cohesion value (c = 31.99 MPa), and a more obvious decrease with the increasing plastic shear strain, compared with the samples treated by higher temperatures (c = 24.67 MPa and 25.88 MPa for T = 100 °C and 200 °C, respectively). This should be explained by the more thermally induced cracks in the heat treated samples, which decreased the initial cohesive strength of the specimens. According to Figure 6b, the thermally treated samples by higher temperatures present higher initial friction angles (φ = 38.87 ° for T = 200 °C, compared with φ = 32.54 ° and 33.74 ° for T = 20 °C and 100 °C, respectively), which should also be resulted from the more thermally induced crack surfaces treated by higher temperatures. Based on these observations, a general trend can be concluded that higher treatment temperatures may lead to relatively lower initial values of cohesion (c) and higher initial values of friction angles (φ) for NRG01 granite samples. This phenomenon should be owing to the different amounts of thermally induced cracks inside the rock under the effects of different temperatures.

This section demonstrates the characteristics of mobilized cohesion and friction angle during the failure process of NRG01 granite samples treated by different temperatures. The non-simultaneous mobilization should be considered in the constitutive models when analyzing the stability of the host rock for site selection or design of a HLW disposal repository.

4.2. Mobilized Hoek-Brown Strength Parameters During Failure of NRG01 Granite

As discussed above, the non-simultaneous mobilization of strength parameters mainly focused on the linear Mohr-Coulomb criterion. For the widely used non-linear Hoek-Brown criterion, only simultaneous mobilization of strength parameters (m and s) can be found to be considered in the published studies [23–26]. It is quite necessary to research the mobilization behaviors of Hoek-Brown strength parameters during the failure process of rock. This section will present such a study based on the laboratory experiments on NRG01 granite samples treated by different temperatures.

4.2.1. Analytical Method

The similar method as illustrated in Figure 4 is used here for determining a series of plastic strain and axial stress values. So the same results presented in Figure 5 can be used in this part of analyses. For each certain plastic shear strain value, the set of stress values under different confining pressures are used to fit the Hoek-Brown criterion [26,42]:

$$\sigma_1 = \sigma_3 + \sigma_{ci}\left(m_b\frac{\sigma_3}{\sigma_{ci}} + s\right)^a, \tag{7}$$

where, σ_1 and σ_3 are the maximum and minimum principal stresses, respectively; σ_{ci} is the uniaxial compression strength of the intact rock; m_b, s and a are the constants for the damaged rock specimens.

It should be noted that the parameter s is related to the fracturing degree of the samples. When σ_3 = 0, the uniaxial compression strength of the damaged samples can be obtained as:

$$\sigma_c = \sigma_{ci}s^a, \tag{8}$$

As the constant a is defined as:

$$a = \frac{1}{2} + \frac{1}{6}\left(e^{-GSI/15} - e^{-20/3}\right), \tag{9}$$

where, *GSI* is the Geological Strength Index indicating the rock quality with the number ranging from 0 (totally fractured) to 100 (intact) [41,42]. It can be seen that for 25 < *GSI* < 100, the value of a is very close to 0.5. Therefore, a is reasonable enough to be set as a constant number 0.5 for simplicity in this study. Thereafter, for the damaged samples at each plastic shear strain, s can be identified based on the corresponding uniaxial compression strength values presented in Figure 5 by the following equation:

$$s = (\sigma_c/\sigma_{ci})^{0.5}, \tag{10}$$

With the determined s values, a series of m_b values can be obtained by fitting the data in Figure 5 with Equation (7). As a result, the mobilized m_b and s values varying with plastic shear strain for NRG01 granite samples treated by different temperatures are presented in Figure 7a,b, respectively.

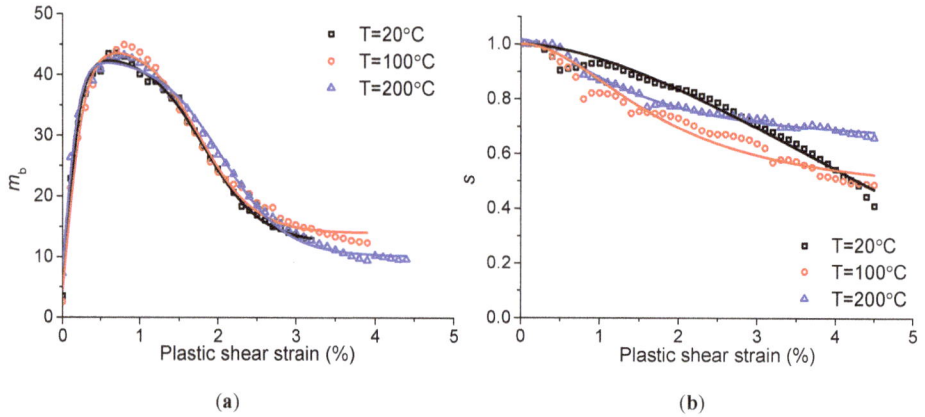

Figure 7. Mobilized Hoek-Brown strength parameters (**a**) m_b and (**b**) s of NRG01 granite treated by different temperatures: T = 20 °C, T = 100 °C and T = 200 °C.

4.2.2. Data Analyses and Discussion

According to Figure 7, for NRG01 granite samples treated by different temperatures (T = 20 °C, 100 °C and 200 °C), several features can be observed as follows:

(1) Similar to the mobilization of cohesion and friction angle, the mobilization of Hoek-Brown strength parameters (m_b and s) is also non-simultaneous during the failure process of NRG01 granite treated by different temperatures no higher than 200 °C;

(2) With increasing plastic shear strain, m_b increases significantly to a maximum value and then decreases until a residual value;

(3) s decreases gradually with the increasing plastic shear strain. This is related to the damage and fracturing process during the tests.

The mobilization of m_b can be fitted with Equation (11):

$$m_b = m_{bi} + (m_{bmax} - m_{bi}) \left[\frac{2}{1 + \exp\left(-5\dfrac{\gamma^p}{\gamma^p_{mb,max}}\right)} - 1 \right]$$
$$- (m_{bmax} - m_{br}) \left[\frac{1}{1 + \exp\left(-5\dfrac{2\gamma^p - \gamma^p_{mb,r}}{\gamma^p_{mb,r}}\right)} \right], \tag{11}$$

where, m_{bi} and m_{br} are the initial and residual value of m_b, respectively; m_{bmax} is the maximum value of m_b; $\gamma^p_{mb,max}$ is the plastic shear strain when m_b is close to its peak value, while $\gamma^p_{mb,r}$ is the plastic shear strain when m_b is close to its residual value.

The mobilization of s can be fitted with Equation (12):

$$s = \frac{s_i - s_r}{1 + \left(\gamma^p / \gamma^p_0\right)^n} + s_r, \tag{12}$$

where, s_i and s_r are the initial and residual value of s, respectively. γ_0^p is the transitional plastic shear strain when the s value turns to decrease in a gradual manner. n is a constant determining the shape of the curve.

The fitted curves are presented in Figure 7, and the fitted coefficients are shown in Table 4. According to Figure 7a, it can be observed that the mobilization of m_b value has very similar characteristics with the increasing plastic shear strain, for the NRG01 granite samples treated by different temperatures (T = 20 °C, 100 °C and 200 °C). Based on a more detailed observation on Figure 7a, we can find that for the NRG01 granite samples treated by the temperature T = 200 °C, the initial value of $m_b = 7.24$. This is apparently higher than the value for the cases of lower temperatures ($m_b = 3.54$ and 2.59, for T = 20 °C and 100 °C, respectively). It is always believed that m_b value is more related to the frictional strength in Mohr-Coulomb criterion [29,56]. For the case of T = 200 °C, there should be more crack surfaces induced by the heat in the granite samples, and this should be the reason why the initial values of friction angle and m_b are both higher. According to Figure 7b, the mobilization of s value are also quite similar with the increasing plastic shear strain, for the NRG01 granite samples treated by different temperatures (T = 20 °C, 100 °C and 200 °C). There are not enough evidence to prove how heat treatment influence the mobilization of s value based on this study. More systematic experimental studies should be carried out in order to make clear the characteristics of mobilized s during the failure process of granite treated by different temperatures.

Table 4. The fitted coefficients determining the mobilized Hoek-Brown strength parameters during the failure of NRG01 granite samples under different treatment temperatures.

T (°C) [1]	m_{bi}	m_{br}	m_{bmax}	$\gamma^p_{mb,max}$	$\gamma^p_{mb,r}$	R^2	s_i	s_r	γ_0^p	n	R^2
20	4.64	12.40	43.71	0.56	3.65	0.9947	1.00	−1.00	8.15	1.72	0.9751
100	4.68	13.98	45.64	0.72	3.53	0.9868	1.00	0.45	1.80	2.00	0.9464
200	8.54	4.09	43.08	0.52	4.09	0.9940	1.00	0.63	1.41	1.70	0.9860

[1] T means the treatment temperature.

5. Conclusions

NRG01 granite samples cored from Alxa candidate area for HLW disposal were treated by different temperatures, and then were used to carry out a series of uniaxial and tri-axial compression experiments under different confining pressures. Complete axial stress—axial strain curves and axial stress—lateral strain curves were recorded. These data were collected to study the mobilization of both Mohr-Coulomb and Hoek-Brown strength parameters during the damage and failure of NRG01 granites samples considering the effect of heat induced by the nuclear waste. According to the analyses in this study, several conclusions can be drawn as follows:

(1) Cohesion weakening and friction angle strengthening occurs during the damage and failure process of NRG01 granite samples treated by different temperatures. However, compared with the findings in the previous studies, cohesion decreases in a more gradual manner for NRG01 granite samples, and the friction angle increases immediately to its maximum value before the cohesion approaching to the residual value. This may be owing to the grain size, heterogeneity, or even the hoop effect induced by the cylindrical shape of the samples. More systematic studies are required to make clear the exact influencing factors, as well as the mechanism.

(2) The temperatures of no higher than 200 °C do not have significant influence on the characteristics of mobilized cohesion or friction angle during the damage and failure process of NRG01 granite samples. However, the samples under room temperature (20 °C) have higher initial cohesion than the samples treated by higher temperatures (T=100 °C and 200 °C). In addition, the samples treated by temperature of 200 °C have higher friction angle than the samples treated by lower temperatures. This should be caused by the cracks induced by the heat treatment.

(3)   The Hoek-Brown strength parameters m_b and s are also observed to show non-simultaneous mobilization behaviors during the failure process of NRG01 granite samples treated by different temperatures. It is found that m_b increases significantly to a maximum value and then decreases until a residual value, and s decreases gradually with the increasing plastic shear strain. The general characteristics of the mobilized m_b and s are similar for NRG01 granite samples treated by different temperatures, and the fitted equations for modelling the mobilization of both parameters are proposed. The samples treated by temperature of 200 °C have higher initial m_b value, this should also be caused by the cracks induced by the heat treatment.

These findings on the mobilization of strength parameters provide a better understanding on the strength properties of NRG01 granite samples, and can be used for building a plastic constitutive model in the next step. This study should also be helpful for guiding the selection and design of HLW disposal repository in Alxa area in China. This study put forward the research on non-simultaneous mobilization of strength parameters to Hoek-Brown strength criterion, and more experimental studies are required to consolidate the results. The methods used in this paper can also be used for this kind of analyses in the other candidate areas for HLW disposal.

Author Contributions: Conceptualization, C.C.; methodology, C.C. and N.X.; formal analysis, C.C. and N.X.; investigation, C.C. and B.Z.; data curation, C.C. and B.Z.; writing—original draft preparation, C.C.; writing—review and editing, C.C.

Funding: This research was funded by Fundamental Research Funds for the Central Universities (Grant No. 2-9-2018-087); the National Natural Science Foundation of China (Grant No. 41772326); and the High-Level Radioactive Waste Disposal Project of the State Administration for Science, Technology and Industry for National Defense, China. The APC was funded by the National Natural Science Foundation of China (Grant No. 41772326).

Acknowledgments: X. Li and S. Li from Institute of Geology and Geophysics, Chinese Academy of Sciences is appreciated here for his support for this work. Y. Dong is also appreciated for his help in this study. The anonymous reviewers gave very helpful suggestions, which were valuable for improving our manuscript.

Conflicts of Interest: The authors declare no conflict of interest.

References

1.   Cheng, C.; Li, X.; Li, S.; Zheng, B. Geomechanical studies on granite intrusions in alxa area for high-level radioactive waste disposal. *Sustainability* **2016**, *8*, 1329. [CrossRef]

2.   Cheng, C.; Li, X.; Li, S.; Zheng, B. Failure behavior of granite affected by confinement and water pressure and its influence on the seepage behavior by laboratory experiments. *Materials* **2017**, *10*, 798. [CrossRef] [PubMed]

3.   Zhao, X.G.; Wang, J.; Cai, M.; Ma, L.K.; Zong, Z.H.; Wang, X.Y.; Su, R.; Chen, W.M.; Zhao, H.G.; Chen, Q.C.; et al. In-situ stress measurements and regional stress field assessment of the Beishan Area, China. *Eng. Geol.* **2013**, *163*, 26–40. [CrossRef]

4.   Zhao, X.G.; Wang, J.; Qin, X.H.; Cai, M.; Su, R.; He, J.G.; Zong, Z.H.; Ma, L.K.; Ji, R.L.; Zhang, M.; et al. In-situ stress measurements and regional stress field assessment in the Xinjiang candidate area for China's HLW disposal. *Eng. Geol.* **2015**, *197*, 42–56. [CrossRef]

5.   Martin, C.D.; Read, R.S.; Martino, J.B. Observations of brittle failure around a circular test tunnel. *Int. J. Rock Mech. Min. Sci.* **1997**, *34*, 1065–1073. [CrossRef]

6.   Read, R.S. 20 years of excavation response studies at AECL's underground research laboratory. *Int. J. Rock Mech. Min. Sci.* **2004**, *41*, 1251–1275. [CrossRef]

7.   IAEA. *Geological Disposal of Radioactive Waste: Technological Implications for Retrievability*; IAEA Nuclear Energy Series No. NW-T-1.19; International Atomic Energy Agency: Vienna, Austria, 2009. Available online: http://www-pub.iaea.org/MTCD/publications/PDF/Pub1378_web.pdf (accessed on 30 September 2019).

8.   Pusch, R. *Geological Storage of Highly Radioactive Waste*; Springer: Berlin/Heidelberg, Germany, 2009.

9.   Hudson, J.A. Underground radio active waste disposal: The rock mechanics contribution. In *ISRM International Symposium—6th Asian Rock Mechanics Symposium*; Sharma, K.G., Ramamurthy, T., Kanjlia, V.K., Gupta, A.C., Eds.; International Society for Rock Mechanics: New Delhi, India, 2010; pp. 3–20.

10. Wang, J. High-level radioactive waste disposal in china: Update 2010. *J. Rock Mech. Geotech. Eng.* **2010**, *2*, 1–11.

11. Chijimatsu, M.; Nguyen, T.S.; Jing, L.; De Jonge, J.; Kohlmeier, M.; Millard, A.; Rejeb, A.; Rutqvist, J.; Souley, M.; Sugita, Y. Numerical study of the thm effects on the near-field safety of a hypothetical nuclear waste repository—bmt1 of the decovalex iii project. Part 1: Conceptualization and characterization of the problems and summary of results. *Int. J. Rock Mech. Min. Sci.* **2005**, *42*, 720–730. [CrossRef]

12. Bond, A.N.C.; Fedors, R.; Lang, P.; McDermott, C.; Neretnieks, I.; Pan, P.; Šembera, J.; Watanabe, N.; Yasuhara, H. Coupled THMC modelling of a single fracture in novaculite for DECOVALEX-2015. In Proceedings of the DFNE2014, Vancouver, BC, Canada, 19–22 October 2014.

13. Chen, L.; Wang, C.P.; Liu, J.F.; Liu, J.; Wang, J.; Jia, Y.; Shao, J.F. Damage and plastic deformation modeling of beishan granite under compressive stress conditions. *Rock Mech. Rock Eng.* **2015**, *48*, 1623–1633. [CrossRef]

14. Zhao, X.G.; Wang, J.; Chen, F.; Li, P.F.; Ma, L.K.; Xie, J.L.; Liu, Y.M. Experimental investigations on the thermal conductivity characteristics of Beishan granitic rocks for China's HLW disposal. *Tectonophysics* **2016**, *683*, 124–137. [CrossRef]

15. Zhao, X.G.; Zhao, Z.; Guo, Z.; Cai, M.; Li, X.; Li, P.F.; Chen, L.; Wang, J. Influence of thermal treatment on the thermal conductivity of Beishan granite. *Rock Mech. Rock Eng.* **2018**, *51*, 2055–2074. [CrossRef]

16. Xue, D.; Zhou, H.; Zhao, Y.; Zhang, L.; Deng, L.; Wang, X. Real-time sem observation of mesoscale failures under thermal-mechanical coupling sequences in granite. *Int. J. Rock Mech. Min. Sci.* **2018**, *112*, 35–46. [CrossRef]

17. Chen, L.; Wang, J.; Zong, Z.; Liu, J.; Su, R.; Guo, Y.; Jin, Y.; Chen, W.; Ji, R.; Zhao, H. A new rock mass classification system Q_{HLW} for high-level radioactive waste disposal. *Eng. Geol.* **2015**, *190*, 33–51. [CrossRef]

18. Kim, J.-S.; Kwon, S.-K.; Sanchez, M.; Cho, G.-C. Geological storage of high level nuclear waste. *KSCE J. Civ. Eng.* **2011**, *15*, 721–737. [CrossRef]

19. Zhao, X.G.; Wang, J.; Cai, M.; Cheng, C.; Ma, L.K.; Su, R.; Zhao, F.; Li, D.J. Influence of unloading rate on the strainburst characteristics of Beishan granite under true-triaxial unloading conditions. *Rock Mech. Rock Eng.* **2013**, *47*, 467–483. [CrossRef]

20. Wang, J. On area-specific underground research laboratory for geological disposal of high-level radioactive waste in China. *J. Rock Mech. Geotech. Eng.* **2014**, *6*, 99–104. [CrossRef]

21. Marschall, P.; Giger, S.; De La Vassière, R.; Shao, H.; Leung, H.; Nussbaum, C.; Trick, T.; Lanyon, B.; Senger, R.; Lisjak, A.; et al. Hydro-mechanical evolution of the edz as transport path for radionuclides and gas: Insights from the Mont Terri rock laboratory (Switzerland). *Swiss J. Geosci.* **2017**, *110*, 173–194. [CrossRef]

22. Wang, J.; Chen, L.; Su, R.; Zhao, X. The Beishan underground research laboratory for geological disposal of high-level radioactive waste in china: Planning, site selection, site characterization and in situ tests. *J. Rock Mech. Geotech. Eng.* **2018**, *10*, 411–435. [CrossRef]

23. Alonso, E.; Alejano, L.R.; Varas, F.; Fdez-Manin, G.; Carranza-Torres, C. Ground response curves for rock masses exhibiting strain-softening behaviour. *Int. J. Numer. Anal. Methods Geomech.* **2003**, *27*, 1153–1185. [CrossRef]

24. Lee, Y.-K.; Pietruszczak, S. A new numerical procedure for elasto-plastic analysis of a circular opening excavated in a strain-softening rock mass. *Tunn. Undergr. Space Technol.* **2008**, *23*, 588–599. [CrossRef]

25. Jianxin, H.; Shucai, L.; Shuchen, L.; Lei, W. Post-peak stress-strain relationship of rock mass based on hoek-brown strength criterion. *Procedia Earth Planet. Sci.* **2012**, *5*, 289–293. [CrossRef]

26. Hajiabdolmajid, V.; Kaiser, P.K.; Martin, C.D. Modelling brittle failure of rock. *Int. J. Rock Mech. Min. Sci.* **2002**, *39*, 731–741. [CrossRef]

27. Zhao, X.G.; Cai, M. A mobilized dilation angle model for rocks. *Int. J. Rock Mech. Min. Sci.* **2010**, *47*, 368–384. [CrossRef]

28. Vermeer, P.A.; De Borst, R. Non-associated plasticity for soils, concrete and rock. *HERON* **1984**, *29*, 163–196.

29. Hajiabdolmajid, V.R. Mobilization of strength in brittle failure of rock. *Géotechnique* **2003**, *53*, 327–336. [CrossRef]

30. Guo, S.; Qi, S.; Zhan, Z.; Zheng, B. Plastic-strain-dependent strength model to simulate the cracking process of brittle rocks with an existing non-persistent joint. *Eng. Geol.* **2017**, *231*, 114–125. [CrossRef]

31. Rafiei Renani, H.; Martin, C.D. Cohesion degradation and friction mobilization in brittle failure of rocks. *Int. J. Rock Mech. Min. Sci.* **2018**, *106*, 1–13. [CrossRef]

32. Schmertmann, J.H.; Osterberg, J.O. *An Experimental Study of the Development of Cohesion and Friction with Axial Strain in Saturated Cohesive Soils*; Research Conference on Shear Strength of Cohesive Soils; ASCE: Reston, VA, USA, 1960; pp. 643–694.

33. Walton, G.; Arzúa, J.; Alejano, L.R.; Diederichs, M.S. A laboratory-testing-based study on the strength, deformability, and dilatancy of carbonate rocks at low confinement. *Rock Mech. Rock Eng.* **2015**, *48*, 941–958. [CrossRef]

34. Martin, C.D. *The Strength of Massive Lac du Bonnet Grantie Around Underground Openings*; University of Manitoba: Winnipeg, MB, Canada, 1993.

35. Martin, C.D.; Chandler, N.A. The progressive fracture of lac du bonnet granite. *Int. J. Rock Mech. Min. Sci. Geomech. Abstr.* **1994**, *31*, 643–659. [CrossRef]

36. Hajiabdolmajid, V.R. *Mobilization of Strength in Brittle Failure of Rock*; Queen's University: Kingston, ON, Canada, 2001.

37. Edelbro, C. Numerical modelling of observed fallouts in hard rock masses using an instantaneous cohesion-softening friction-hardening model. *Tunn. Undergr. Space Technol.* **2009**, *24*, 398–409. [CrossRef]

38. Barton, N.; Pandey, S.K. Numerical modelling of two stoping methods in two indian mines using degradation of c and mobilization of φ based on q-parameters. *Int. J. Rock Mech. Min. Sci.* **2011**, *48*, 1095–1112. [CrossRef]

39. Walton, G.; Diederichs, M.; Punkkinen, A.; Whitmore, J. Back analysis of a pillar monitoring experiment at 2.4km depth in the Sudbury Basin, Canada. *Int. J. Rock Mech. Min. Sci.* **2016**, *85*, 33–51. [CrossRef]

40. Walton, G. Initial guidelines for the selection of input parameters for cohesion-weakening-friction-strengthening (CWFS) analysis of excavations in brittle rock. *Tunn. Undergr. Space Technol.* **2019**, *84*, 189–200. [CrossRef]

41. Hoek, E.; Carranza-Torres, C.; Corkum, B. Hoek-Brown failure criterion. *Proc. NARMS* **2002**, *1*, 267–273.

42. Hoek, E.; Brown, E.T. The Hoek—Brown Failure Criterion and GSI. *J. Rock Mech. Geotech. Eng.* **2019**, *11*, 445–463. [CrossRef]

43. David, E.C.; Brantut, N.; Schubnel, A.; Zimmerman, R.W. Sliding crack model for nonlinearity and hysteresis in the uniaxial stress–strain curve of rock. *Int. J. Rock Mech. Min. Sci.* **2012**, *52*, 9–17. [CrossRef]

44. Zhao, X.G.; Cai, M.; Wang, J.; Li, P.F. Strength comparison between cylindrical and prism specimens of Beishan granite under uniaxial compression. *Int. J. Rock Mech. Min. Sci.* **2015**, *76*, 10–17. [CrossRef]

45. Song, F.; Dong, Y.-H.; Xu, Z.-F.; Zhou, P.-P.; Wang, L.-H.; Tong, S.-Q.; Duan, R.-Q. Granite microcracks: Structure and connectivity at different depths. *J. Asian Earth Sci.* **2016**, *124*, 156–168. [CrossRef]

46. Shi, X. The Tectonic Affinity of the Zongnaishan-Shalazhashan Zone in Northern Alxa and Its Implications: Evidence from Intrusive and Metamorphic Rocks. Ph.D. Thesis, Chinese Academy of Geological Sciences, Beijing, China, 2015. (In Chinese).

47. Zhou, J.; Lan, H.; Zhang, L.; Yang, D.; Song, J.; Wang, S. Novel grain-based model for simulation of brittle failure of Alxa porphyritic granite. *Eng. Geol.* **2019**, *251*, 100–114. [CrossRef]

48. Chen, L.; Liu, J.; Wang, C.; Wang, X.; Su, R.; Wang, J.; Shao, J. Elastoplastic damage model of beishan deep granite. *Yanshilixue Yu Gongcheng Xuebao/Chin. J. Rock Mech. Eng.* **2013**, *32*, 289–298.

49. Guo, R. *Coupled Thermal-Mechanical Modelling of a Deep Geological Repository Using the Horizontal Tunnel Placement Method in Sedimentary Rock Using Code_Bright*; Report NWMO TR-2010-22; Nuclear Waste Management Organization: Toronto, ON, Canada, 2010; p. 54. Available online: https://www.nwmo.ca/~{|}/media/Site/Reports/2015/09/24/05/33/1695_nwmotr-2010-22horizontal-tunnel-placmentmethod_limestone_r0b_worddocument.ashx?la=en (accessed on 30 September 2019).

50. Ikonen, K. *Thermal Analyses of Spent Nuclear Fuel Repository*; POSIVA 2003-04; Posiva: Eurajoki, Finland, 2003; p. 61. Available online: http://www.posiva.fi/files/1018/Posiva_2003-04.pdf (accessed on 30 September 2019).

51. Rutqvist, J.; Chijimatsu, M.; Jing, L.; Millard, A.; Nguyen, T.S.; Rejeb, A.; Sugita, Y.; Tsang, C.F. A numerical study of thm effects on the near-field safety of a hypothetical nuclear waste repository—bmt1 of the decovalex iii project. Part 3: Effects of thm coupling in sparsely fractured rocks. *Int. J. Rock Mech. Min. Sci.* **2005**, *42*, 745–755. [CrossRef]

52. Weetjens, E. *Update of the Near Field Temperature Evolution Calculations for Disposal of UNE-55, MOX-50 and Vitrified HLW in a Supercontainer-Based Geological Repository*; SCK·CEN-ER-86; Belgian Nuclear Research Centre: Mol, Belgium, 2009; p. 14.

53. Gajewski, M.; Jemioło, S. The influence of pavement degradation caused by cyclic loading on its failure mechanisms. *Balt. J. Road Bridge Eng.* **2016**, *11*, 179–187. [CrossRef]

54. Walton, G.; Diederichs, M.S. A new model for the dilation of brittle rocks based on laboratory compression test data with separate treatment of dilatancy mobilization and decay. *Geotech. Geol. Eng.* **2015**, *33*, 661–679. [CrossRef]

55. Walton, G.; Alejano, L.R.; Arzua, J.; Markley, T. Crack damage parameters and dilatancy of artificially jointed granite samples under triaxial compression. *Rock Mech. Rock Eng.* **2018**, *51*, 1637–1656. [CrossRef]

56. Hoek, E. Strength of jointed rock masses. *Geotechnique* **1983**, *33*, 187–223. [CrossRef]

MDPI

St. Alban-Anlage 66

4052 Basel

Switzerland

Tel. +41 61 683 77 34

Fax +41 61 302 89 18

www.mdpi.com

Energies Editorial Office

E-mail: energies@mdpi.com

www.mdpi.com/journal/energies

www.ingramcontent.com/pod-product-compliance
Lightning Source LLC
Chambersburg PA
CBHW051916210326
41597CB00033B/6158